NEUROPEPTIDE Y

MOLECULAR STRUCTURE, ROLE IN FOOD INTAKE AND DIRECT/INDIRECT EFFECTS

PROTEIN BIOCHEMISTRY, SYNTHESIS, STRUCTURE AND CELLULAR FUNCTIONS

Additional books in this series can be found on Nova's website
under the Series tab

Additional e-books in this series can be found on Nova's website
under the e-book tab

NEUROSCIENCE RESEARCH PROGRESS

Additional books in this series can be found on Nova's website
under the Series tab.

Additional e-books in this series can be found on Nova's website
under the e-book tab

PROTEIN BIOCHEMISTRY, SYNTHESIS, STRUCTURE AND CELLULAR FUNCTIONS

NEUROPEPTIDE Y

MOLECULAR STRUCTURE, ROLE IN FOOD INTAKE AND DIRECT/INDIRECT EFFECTS

STEVEN L. PARKER
EDITOR

New York

NOTICE TO THE READER

Library of Congress Cataloging-in-Publication Data

Library of Congress Control Number: 2013934500

ISBN: 978-1-62618-421-3

Published by Nova Science Publishers, Inc. † *New York*

Contents

Preface

This book presents a broad summary of current knowledge concerning the structure and biochemistry of neuropeptide Y in relation to its role in feeding and interactions with receptors, transducers, effectors and channels that can also connect to food intake. The very high conservation of NPY sequence indicates its critical importance in basal metabolic regulation. Mutations connecting to NPY are virtually absent, and those affecting the Y1 and Y2 receptors are very few. From the evidence presented across vertebrate classes, NPY appears mainly as metabotropic driver via the Y1 group of receptors, and its negative metabotropicity through the Y2 receptor is only of importance in the mammal, and then tempered by Y2 receptor masking in the hypothalamus and low numbers in the cortex, and by low availability of NPY outside the forebrain. Anatomically, NPY is well-represented especially in limbic areas of the forebrain, but the feeding-critical presence of Y1 receptors is, at least in the rodent, highest in the neocortex, and evidence presented in this book points to much larger involvement of cortical NPY receptors in feeding regulation than is usually perceived. The strong presence of this highly conserved peptide in the vertebrate forebrain is increasingly documented as linked not only to the regulation of glucose metabolism (and insulin activity), but also as directly involved in the operation of transducers, effectors and channels. This should be enabled especially by the high interactivity of the acidic sector of NPY (which is not shared by NPY-related systemic peptides, peptide YY and pancreatic polypeptide), and results in partial agonism and possibly also in accumulation in the bilayer. NPY could be involved as a helper agonist in several parts of the vertebrate metabolome.

Chapter I – Since the presence of neuropeptide Y (NPY) in the central nervous system of chickens was demonstrated in 1988, numerous studies have shown that NPY functions as a strong orexigenic factor in this species. Similar to mammals, central administration of NPY robustly stimulates ingestive behavior in chickens in a dose-dependent manner. NPY is one of the most abundant peptides in chicken brain. Fasting leads to increased NPY gene expression in this organ. Six receptor subtypes for NPY (Y1, Y2, Y4, Y5, Y6 and Y7) have been identified in chickens and their binding characteristics have been investigated. Although pharmacological studies prove that appetite stimulation by NPY in mammals is mediated by receptors Y1 and Y5, there are conflicting data between the receptor expressions and pharmacological effects in chickens. This review provides an overview of the orexigenic effect of NPY and interrelationship with appetite-related signals, and summarizes the role of the NPYergic system on feeding regulation in chickens.

Chapter II – Obesity is a public health problem worldwide, affecting both developed societies and developing countries. The central nervous system has evolved a meticulously interconnected circuitry in order to maintain energy homeostasis and adequate nutritional state. Neuropeptide Y (NPY), a 36-amino-acid peptide, is one of the most abundant and widely distributed peptides in the neural matrix. Among physiological actions of NPY, potent orexigenic effects have possibly the largest impact. NPY and its receptors have been implicated in a variety of physiological effects such as regulation of food intake, energy balance, hormone release, cardiovascular disease, thermoregulation, stress response and anxiety. There are at least four receptors commonly expressed across the vertebrates that specifically respond to NPY and its analogs. These include the structurally quite similar Y1, Y4 and Y5 subtypes and the rather divergent Y2 receptor. This complex system is embedded in a densely redundant network which helps stable energy homeostasis. Previously the authors showed that NPY increases in a compensatory manner during weight loss in obese adolescents. This review focuses on effects of NPY in obesity and the metabolic syndrome. The authors also examine the neuroendocrine regulation of food intake.

Chapter III – Neuropeptide Y (NPY) is an orexigenic peptide of 36-amino acids originally isolated from porcine brain. NPY and family peptides, peptide YY and pancreatic polypeptide, have pleiotropic effects upon food intake and energy homeostasis through NPY receptor family. Among known 8 types of NPY receptors, Y1, Y2 and Y5 receptors are the most studied in humans. A lot of experimental and clinical evidence from in vivo studies using selective NPY receptor agonists or antagonists, and from genetic manipulation, suggests that Y1 and Y5 receptors are involved in orexigenic signals and that Y2 receptor contributes anorexigenic signals. NPY neurons in hypothalamic arcuate nucleus not only project to ventromedial hypothalamus and inhibit satiety, but also project to lateral hypothalamus and increase food intake. Despite significant effects of NPY on food intake as well as energy homeostasis by pharmacological manipulations, knockdown of NPY gene does not result in dramatic changes of phenotypic obesity. NPY effects on energy balance may be redundant due to contributions of central signals especially by melanocortin and orexin, or of a variety of peripheral signals, including particularly those by leptin, ghrelin and insulin. On the other hand, the NPY Y2 receptor might be involved in visceral obesity by direct effects on adipose tissues under stress with high-caloric diet, and in the development of plaque instability by direct effects on vascular smooth muscle cells. Associations between NPY/NPY receptor single nucleotide polymorphisms (SNPs) and obesity/type 2 diabetes mellitus / dyslipidemia / atherosclerosis have been reported in some cohort studies. The authors have reported that Y2R SNPs were associated with serum high density lipoprotein cholesterol levels. This review focuses on recent evidence about the pathophysiological mechanisms of NPY action, and on evaluation of NPY and NPY receptor promise as prophylactic and therapeutic targets in metabolic disorders and atherosclerosis-related diseases.

Chapter IV – Neuropeptide Y (NPY) is a 36-amino acid peptide widely distributed in both the central and peripheral nervous systems. NPY and its receptors play extremely diverse roles in the nervous system including regulation blood pressure, circadian rhythms, feeding behavior, anxiety, vasoconstriction and gastrointestinal tract motility. In mammals, NPY has been revealed in the majority of sympathetic ganglion neurons, in a high number of neurons of parasympathetic cranial ganglia as well as of intramural ganglia. NPY exerts its action via specific receptors, designated as Y1 to Y6, of which Y1, Y2, Y4, Y5 and Y6 have been cloned. All known NPY receptors belong to family A (rhodopsin-like) of the large superfamily of G-protein-coupled heptahelical receptors. Actions of NPY on peripheral target-organs are predominantly realized through postsynaptic receptors Y1, Y3–Y5, and presynaptic Y2 receptors. NPY is present in large dense-cored vesicles and is released at high-frequency stimulation. NPY affects not only vascular tone, frequency and strength of heart contractions, motorics and secretion of the gastrointestinal tract, but also has trophic effect and produces proliferation of cells of organs-targets, specifically of vessels, myocardium, and adipose tissue. During early postnatal development, the percentage of the NPY-containing neurons in many autonomic ganglia increases. In aged organisms, the proportion of NPY-immunopositive neurons decreases. This seems to be connected with the trophic NPY effect on target cells, as well as with regulation of their functional state.

Chapter V – Neuropeptide Y (NPY) participates in the regulation of aminergic transmission, glucose homeostasis, secretion of hormonal and other peptides, and activity of channels and transporters. Connected to these processes, NPY affects conditions related to feeding, hypertension, diabetes and angiogenesis. These multiple activities are supported by two cooperating NPY receptors, the Y1 and Y5 subtypes, and by the Y2 receptor (possibly in association with the Y4 receptor) somewhat opposing the Y1 and Y5 (especially in the mammal). This counteraction is aided by the proteolytic clipping of Y1/Y5/ Y2-active 36-residue Y peptides to Y2-specific 3-36 peptides. This chapter re-examines the properties of brain NPY receptors and presents new evidence for receptor subtype- and brain area-related differences in transduction via Y1 and Y2 receptors.

The broad, protean interactions of the 10-16 EDAPAED section of NPY could be important in the observed blockade of the Y1 receptor by NPY and interactions with transducers and effectors (including the voltage sensors of ion channels).

The homobasic 19-36 sector of NPY could interact with extracellular domains of channels. The G-protein-coupling intracellular domains of Y receptors could also couple with effectors, including transporters and channels.

With NPY as ligand (but much less with the related peptide YY), the seesaw feeding regulation via Y1/Y5 vs. Y2/(Y4) receptors could significantly depend on these additional interactions.

In: Neuropeptide Y
Editors: Steven L. Parker

ISBN: 978-1-62618-421-3
© 2013 Nova Science Publishers, Inc.

Chapter I

Chicken Neuropeptide Y in the Control of Appetite and Metabolism

Takashi Bungo[1,2][*], *Jun-ichi Shiraishi*[3], *Shin-ichi Kawakami*[1], *D. Michael Denbow*[4] *and Mitsuhiro Furuse*[5]

[1]Laboratory of Animal Behavior and Physiology, Graduate School of Biosphere Science, Hiroshima University, Higashi-Hiroshima, Japan
[2]Japanese Avian Bioresource Project Research Center, Hiroshima University, Higashi-Hiroshima, Japan
[3]Department of Animal Science, Nippon Veterinary and Life Science University, Tokyo, Japan
[4]Department of Animal and Poultry Sciences, Virginia Polytechnic Institute and State University, Blacksburg, VA, US
[5]Laboratory of Regulation in Metabolism and Behavior, Faculty of Agriculture, Kyushu University, Fukuoka, Japan

[*] Correspondence should be addressed to: Dr. T. Bungo, E-mail: bungo@hiroshima-u.ac.jp

Abstract

Since the presence of neuropeptide Y (NPY) in the central nervous system of chickens was demonstrated in 1988, numerous studies have shown that NPY functions as a strong orexigenic factor in this species. Similar to mammals, central administration of NPY robustly stimulates ingestive behavior in chickens in a dose-dependent manner. NPY is one of the most abundant peptides in chicken brain. Fasting leads to increased NPY gene expression in this organ. Six receptor subtypes for NPY (Y1, Y2, Y4, Y5, Y6 and Y7) have been identified in chickens and their binding characteristics have been investigated. Although pharmacological studies prove that appetite stimulation by NPY in mammals is mediated by receptors Y1 and Y5, there are conflicting data between the receptor expressions and pharmacological effects in chickens. This review provides an overview of the orexigenic effect of NPY and interrelationship with appetite-related signals, and summarizes the role of the NPYergic system on feeding regulation in chickens.

Introduction

Neuropeptide Y (NPY), a 36-amino acid neurotransmitter that belongs to the pancreatic polypeptide family (Tatemoto et al., 1982), activates multiple NPY receptors. NPY neurons abundantly innervate the hypothalamus, where NPY is involved in the regulation and integration of a broad range of homeostatic functions (Allen et al., 1983). To date six G-protein coupled NPY receptors have been identified in chickens; their characteristics are discussed in detail in this chapter. Effects of NPY on feeding regulation have been widely investigated since this neuropeptide was discovered 30 years ago. The hypothalamus plays a central role in regulating appetite and metabolism. However, networks within the hypothalamus that regulate food intake and metabolism, and the effects of fasting on those pathways are not completely understood in any animal. Appetite control in chickens is one of the greatest concerns of animal scientists and producers because it not only is important for general health, but also affects productivity for meat and eggs. Thus, understanding the regulation of appetite is a very important theme in poultry industry. In this chapter we present a review of literature showing how NPY integrates feeding behavior, including energy metabolism in chickens.

Chicken NPY in the Hypothalamus

Kuenzel (1989) suggests one neural pathway play a direct role in the control of feeding behavior in the chicken. The autonomic nervous system, which plays a direct role in the feeding regulation, contains the ventromedial region of the hypothalamus, which is connected to the sympathetic nervous system and the paraventricular nucleus (PVN), which has connections both to the parasympathetic and sympathetic components of the autonomic nervous system. The observation of NPY gene expression in brain regions involved in appetite regulation is consistent with the recognized importance of NPY in feeding regulation in a variety of species, including the chicken (Leibowitz, 1989). We therefore briefly summarize findings about the hypothalamic NPY.

The presence of NPY in the central nervous system of chickens was first demonstrated by immunohistochemistry using antibodies against porcine NPY (Kuenzel and McMurtry, 1988). Later studies confirmed the distributions of NPY-immunoreactive cells and fibers in chick hypothalamus observed by Kuenzel and McMurtry (1988). Esposito et al. (2001) found NPY-immunoreactive neurons in the nucleus periventricularis hypothalami (PHN), the PVN, the regio lateralis hypothalami (LHy), the nucleus infundibuli (IN), and the zona interna of the median eminence (ME). Recent study of immunohistochemistry in the IN also revealed the presence of NPY neurons (Shiraishi et al., 2011a). Collectively, NPY-immunoreactive fibers were seen throughout the chick brain, but were more abundant in the hypothalamus where they formed networks and pathways. The brain parasagittal sections exhibited two NPY-immunoreactive fiber pathways: (1) in the ventral hypothalamus, from the preoptic area just above the optic chiasm to the ME; and (2) in the dorsal hypothalamus, from the PVN region to the IN region. The coronal sections also showed a fiber pathway running from the IN toward the ME.

The results of *in vitro* autoradiography and *in situ* hybridization showed a very similar distribution pattern with immunohistochemical studies mentioned above (Merckaert and Vandesande, 1996; Boswell et al., 1998; Wang et al., 2001). While the intensity of labeling varied among nuclei, NPY gene expression was widely distributed in the hypothalamus: PVN, the nucleus ventromedialis hypothalami (VMN) at the base of the third ventricle, the IN within the mammillary hypothalamic region, the nucleus preopticus periventricularis around the third ventricle, the nucleus preopticus medialis, nucleus anterior medialis hypothalami, and the PHN (Wang et al., 2001). Findings indicating that these brain regions are involved in appetite regulation are consistent with the recognized importance of NPY in the control of appetite in a variety of species.

To summarize the aforementioned results, Table 1 shows the comparison of the localization of NPY mRNA with the immunocytochemical localization of NPY fibers, perikarya and neurons in the chicken diencephalon (Kuenzel and McMurtry, 1988; Esposito et al., 2001; Wang et al., 2001; Shiraishi et al., 2011).

Table 1. Comparison of the localization of NPY mRNA with the immnocytochemical localization of NPY fibers, perikarya and neurons in the chicken diencephalon (Kuenzel and McMurtry, 1988; Esposito et al., 2001; Wang et al., 2001; Shiraishi et al., 2011)

	NPY mRNA labeling	NPY-ir fibers/perikarya/neurons
GL	+	++
SL	+++	+
SM	++	+
HM	+	+
DMA	+	+
ROT	++++	++
VMN	++	++
PVN	++	+++
POM	+	++
POP	++++	++
AM	+	++
PHN	++	+++
IN	++++	+++
nCPa	++++	++

NPY-ir: NPY immunoreactive.
GLv: nucleus geniculatus lateralis, pars ventralis, SL: nucleus septalis lateralis.
SM: nucleus septalis medialis, HM: nucleus habenularis medialis.
DMA: nucleus dorsomedialis anterior thalami, ROT: nucleus rotundus.
VMN: nucleus ventromedialis hypothalami.
PVN: nucleus paraventricularis magnocellularis, POM: nucleus preopticus medialis.
POP: nucleus preopticus periventricularis, PHN: nucleus periventricularis hypothalami.
AM: nucleus anterior medialis hypothalami, IN: nucleus infundibuli
CPa: commissura pallii
+ low, ++ moderate, +++ dense, ++++ very dense.

Chicken NPY Receptors

NPY receptors are a family of seven transmembrane G-protein coupled receptors, that are expressed throughout the central and peripheral nervous systems where they mediate a variety of responses ranging from regulation of metabolism and food intake to regulation of neurotransmitter release. In all studied mammals, the expressed subtypes are Y1, Y2, Y4 and Y5 (Michel et al., 1998); the Y6 gene is a pseudogene in man and pig, but is active in the rabbit and mouse (Starbäck et al., 2000). The proposed NPY-selective Y3 subtype (Wahlestedt et al., 1992) thus far was not cloned. In chickens, six NPY receptors (Y1, Y2, Y4, Y5, Y6 and Y7) have been identified and their binding characteristics have been investigated (Salaneck et al., 2000; Holmberg et al., 2002; Lundell et al., 2002; Bromée et al., 2006). In mammals, several lines of evidence suggest that the NPY Y1 and Y5 receptors, alone or together, could mediate the action of NPY on feed intake and metabolism. The genes for these receptors are located on the same chromosome in man (Herzog et al., 1997) and mouse (Nakamura et al., 1997), apparently have similar promoters, could be transcribed simultaneously, and of course would be expressed in the same cells. Thus, these two receptors should qualify as twin receptors mediating the NPY-induced hyperphagia.

In order to understand NPY-mediated behavioral, autonomic and neuroendocrine effects in chickens, it is important to characterize in detail the distribution of the central NPY receptors. The open reading frame of the chicken NPY Y1 receptor has 385 amino acids. The chicken Y1 has 80–83% overall sequence identity and 90–92% identity in the transmembrane regions with mammalian Y1 receptors (Holmberg et al., 2002). The NPY Y1 receptor was pharmacologically defined by high affinity for porcine NPY and porcine [Leu31,Pro34]NPY and low affinity for N-terminally truncated NPY analogues (chicken, human and rat pancreatic polypeptide (PP), and porcine ([D-Trp32]NPY) (Holmberg et al., 2002; Table 2). The *in situ* hybridization showed that Y1 mRNA is highly expressed in the IN, which is a chicken structure homologous to the arcuate nucleus (ARC) in mammals (Holmberg et al., 2002).

The isolation of the mammalian NPY Y2 receptor gene by expression cloning was reported in 1995 by two groups (Rose et al., 1995; Gerald et al., 1995). The chicken Y2 receptor displays 75–80% sequence identity to mammalian Y2 receptors. However, the chicken receptor differs significantly from the mammalian by an appreciable affinity for the mammalian Y1-selective agonist porcine [Leu31, Pro34]NPY, and a low affinity for the mammalian Y2-selective antagonist BIIE0246. The rank order of affinities was: chicken peptide YY (PYY), chicken NPY, porcine NPY(2-36), porcine NPY(3-36) > porcine NPY(13-36) > porcine [Leu31, Pro34]NPY > porcine NPY(18-36) > porcine [D-Trp32]NPY, BIIE0246 > rat PP, human PP, chicken PP, BIBP3226, SR120819A (Table 2). The *in situ* hybridization detected Y2 expression in the hippocampus, an area where Y2 is abundant also in mammals (Salaneck et al., 2000).

The NPY Y4 receptor was originally named PP1 receptor, because of its preference for pancreatic polypeptide as a ligand. In mammals, the affinity of PP for this receptor is much greater than the affinities of NPY and PYY (Bard et al., 1995; Lundell et al., 1995, 1996; Gregor et al., 1996). In chickens, the receptor has 377 amino acids (Lundell et al., 2002) and has 57–60% overall amino acid sequence identity to the mammalian Y4R, with 70% identity in the transmembrane regions. The Y4 mRNA was detected in the brainstem, cerebellum and hippocampus (Lundell et al., 2002).

**Table 2. Binding profiles of chicken NPY receptors
(Salaneck et al., 2000; Lundell et al., 2002; Holmberg et al., 2002;
Bromée et al., 2006)**

Subtype	Binding profile of chicken receptors
Y1	porcine NPY, porcine [Leu31, Pro34]NPY > chicken PYY > porcine NPY2-36, porcine NPY3-36, porcine NPY13-36, BIBP3226 >> porcine PP, human PP, chicken PP, porcine [D-Trp32]NPY, BIIE0246, SR120819A
Y2	chicken PYY, chicken NPY, porcine NPY2-36, porcine NPY3-36 > porcine NPY13-36 > porcine [Leu31, Pro34]NPY > porcine NPY18-36 > porcine [D-Trp32]NPY, BIIE0246 > rat PP, human PP, chicken PP, BIBP3226, SR120819A
Y4	porcine PYY, human PP, chicken PYY > porcine [Leu31, Pro34]NPY, chicken PP, porcine NPY, porcine NPY2-36, porcine NPY13-36 > porcine NPY18-36, porcine [D-Trp32]NPY >> BIBP3226
Y5	porcine [Leu31, Pro34]NPY, porcine NPY, CGP71863A, porcine NPY(3-36), porcine NPY(2-36), human PP, chicken PYY, porcine [D-Trp32] NPY, rat PP > porcine NPY(18–36), chicken PP >> BIBP3226, SR120819A
Y6	not detected
Y7	porcine PYY > porcine NPY > chicken PYY > porcine [Leu31, Pro34]NPY, porcine NPY3-36 > porcine NPY13-36 > chicken PP

NPY: neuropeptide Y, PYY: peptide YY, PP: pancreatic polypeptide.

The open reading frame of the chicken Y5 receptor gene codes for 443 amino acids. The full-length chicken NPY Y5 receptor displays 64-72% overall and 85-86% transmembrane identity to mammalian Y5 receptors (Holmberg et al., 2002). The rank order of affinities of the ligands for the chicken Y5 receptor was: porcine [Leu31, Pro34]NPY, porcine NPY, CGP71863A, porcine NPY(3–36), porcine NPY(2–36), human PP, chicken PYY, porcine [D-Trp32] NPY, rat PP > porcine NPY(18–36), chicken PP >> BIBP3226, SR120819A (Table 2). Anatomically, there is high Y5R mRNA expression in the IN (Holmberg et al., 2002). These results indicate that NPY Y1 and Y5 receptors might be "feeding receptors" for the action of NPY on food intake in chickens.

Chicken NPY Y6 receptor was also cloned (Bromée et al., 2006). It has 61–63% amino acid identity to the functional mammalian Y6 receptors. *Y6* is a pseudogene in some mammals (Burkhoff et al., 1998), whereas it seems to remain functional in others. The NPY Y6 receptor mRNA has been detected in gastrointestinal tract, adipose tissue and brain, particularly in the hypothalamus (Bromée et al., 2006). The Y6 receptor clearly does not contribute to the physiological effects of NPY in humans, but must be taken into account when considering physiological effects of NPY in chickens. Functions of the Y6 receptor in the chicken hypothalamus need further investigation especially with respect to the control of appetite.

The NPY Y7 receptor was cloned from the zebrafish and the chicken. This receptor was not found in any mammal thus far. The chicken Y7 receptor has 65% overall amino acid identity to the zebrafish Y7 receptor (Bromée et al., 2006). The identity between chicken Y7 and chicken or mammalian Y2 is 50–55%, the same degree of identity that is observed between zebrafish Y7 and Y2. While the zebrafish Y7R is expressed in brain, eye and intestine (Fredriksson et al., 2004), the chicken Y7R was only detected in adrenal gland (Bromée et al., 2006). The physiological function of the NPY Y7 receptor remains to be clarified.

A more detailed account of the relationship between the NPY receptors and appetite follows.

Central NPY on Energy Homeostasis

The energy stored in the adipose tissue is critical for normal physiology and survival. Energy flux must be monitored and adjusted to ensure that energy intake and expenditure remain within acceptable limits. Feeding of course is critical to energy homeostasis, and there is need of balance between intake and expenditure. Feeding behavior is a set of responses involving several factors that signal the body needs in energy to the central nervous system, particularly to the hypothalamus. The hypothalamus integrates neuronal, metabolic and endocrine signals, and uses different effector pathways for activation of behavioral responses and neuroendocrine axes.

The first step toward identifying NPY as a physiologically important regulator of food intake would be to establish a direct correlation between increase in brain NPY levels and food intake. In the hypothalamus, NPY is synthesized primarily by neurons located in the ARC, and these cells are activated in response to negative energy balance (e.g., caloric restriction or starvation). From the viewpoint of orexigenic effects of NPY, this activation of NPY neurons is proposed to contribute to hyperphagic trigger by reducing the production of glucose (through decrease of cAMP formation (Chance et al., 1989; Aakerlund et al., 1990) and creating a negative energy balance. In the mammal, increased rate of NPY synthesis in the ARC predates NPY release (Kalra et al., 1991; Dube et al., 1992; Sahu et al., 1992). The hypothalamic ARC of mammals is an integration region with a reduced brain-blood barrier and abundant NPY neurons that mediate peripheral and central regulatory signals in ARC-PVN pathway, feeding behavior and energy homeostasis (Schwartz et al., 2000). A negative energy balance resulting from food deprivation, promotes increased NPY synthesis and release in the ARC–PVN axis, and with repletion of energy stores frequency of the episodic NPY secretion rapidly decreases, followed by a gradual normalization of NPY synthesis (Kalra et al., 1991; Sahu et al., 1992). These observations indicate that NPY is a physiological appetite transducer.

Studies in chickens also revealed that depletion in energy stores, as that produced by fasting or food restriction, enhances hypothalamic NPY mRNA expression (Boswell et al., 1999a, b; Wang et al., 2001). The increased NPY content in the IN seems to derive from an elevated rate of NPY synthesis, as suggested by the finding that neural expression of NPY mRNA in the avian IN is activated by fasting (Boswell et al., 2002). Zhou et al. (2005) also found that NPY content of the hypothalamic IN and PVN, but not LHy gradually increased with fasting. Subsequent feeding restored the NPY level in the PVN to the pre-fasting level. Furthermore, recent research using microarray and qRT-PCR analysis revealed that not only the NPY ligand but also the Y5 receptor in the chicken hypothalamus was increased by fasting (Higgins et al., 2010). Both findings support the involvement of the hypothalamic NPYergic system in the regulation of energy homeostasis in chickens.

In mammals, the increased hypothalamic NPY induced by food deprivation or restriction contributes to the regulation of energy balance by promoting weight gain in situations when animals are underweight (Davies and Marks, 1994). Central injection of NPY promotes anabolism by decreasing thermogenesis and stimulating the activity of lipogenic enzymes (Clark et al., 1984; Billington et al., 1994). Thus, NPY mediates a compensatory response to starvation by promoting feeding and energy storage when food becomes available. In chickens, injection of NPY into the brain also induces hypothermia (Tachibana et al., 2004) and hyperinsulinemia (Kuenzel and McMurtry, 1988). Levels of plasma glucose and triacylglycerol were decreased, while those of non-esterified fatty acid increased after NPY administration (Tachibana et al., 2006). Recently, we found that central injection of oligodeoxynucleotides against NPY Y1 receptor induced hyperthermia in chicks (unpublished data). Taken together with these observations and experimental evidence showing that the possibility of a compensatory mechanism in the hypothalamic NPYergic system of chickens, as well as mammals. A summary on the effects of negative energy balance upon the central NPYergic system, and of exogenous NPY signaling in the periphery is depicted in Figure 1.

Central NPY on Appetite

Clark et al. (1984) were the first to demonstrate that intra-cerebroventricular (ICV) injection of NPY causes robust feeding in rats. The ICV injection of NPY has a long-lasting effect, even up to 24 h (Morley et al., 1987). In chickens, Kuenzel et al. (1987) revealed that NPY administered via the lateral ventricle of the brain effects a marked increase in food intake, approximately twice of that observed with control chickens. Substantial evidence indicates that central NPY is one of the most potent stimulators of feeding in chickens (Kuenzel and McMurtry, 1988; Furuse et al., 1997; Bungo et al., 2000; Dodo et al., 2005; Saneyasu et al., 2011).

There is a markedly biphasic effect of NPY on feeding behavior in chicks. Moderate NPY doses induce hyperphagia, while small doses tend to be hypophagic in 2-day-old chicks (Steinman et al., 1987). The latter effect of NPY appears to be inversely related to the increased occurrence of convulsions in chicks. Saneyasu et al. (2011) reported that theY1 mRNA expression at 2 days of age tended to be higher than those at 1, 4 and 8 day in layer chicks.

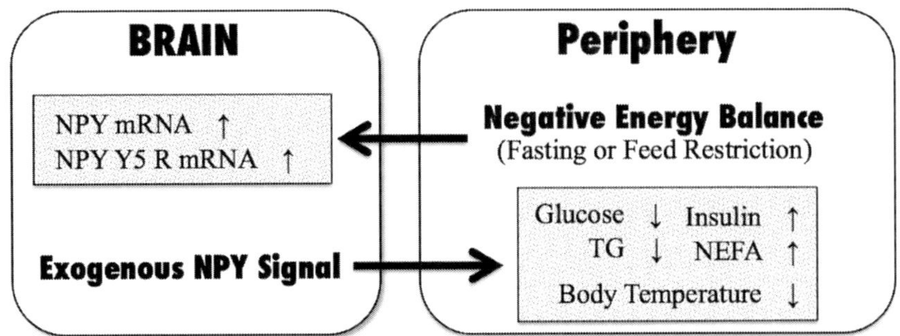

Figure 1. Summarized effects of negative energy upon central NPYergic system balance, and of peripheral effects of exogenous NPY signal (Kuenzel and McMurtry, 1988; Boswell et al., 1999a,b; Wang et al., 2001; Tachibana et al., 2004; Zhou et al., 2005; Tachibana et al., 2006; Higgins et al., 2010). NPY: neuropeptide Y, Y5 R: Y5 receptor, TG: triacylglycerol, NEFA: non-esterified fatty acid.

The stimulation via the NPY Y1 receptor might have relationship with the NPY-induced convulsions, which is age-specific. Also, these changes might be related with the shift of the nutrient source because chicks depend on their yolk sac for the bulk of their nutrition in the first three days of life (Freeman, 1965). They should therefore change the program of the appetite regulatory system at this stage.

NPY has been shown to interact with numerous neurotransmitter and neuropeptide systems that participate in the control of appetite. In mammals, the ARC is the major site of expression for NPY within neurons in the hypothalamus that projects to the PVN, the VMN and other sites, facilitating the interaction of NPY with other orexigenic and anorexigenic signals (Sahu et al., 1988). The ARC contains neuroendocrine neurons, the so called melanocortin system which is an essential mediator for the peripheral signals such as leptin and insulin action in the central nervous system (Schwartz et al., 2000; Benoit et al., 2002). In an expanded sense, the melanocortin system consists of pro-opiomelanocortin (POMC) neurons inhibiting feeding, and NPY and/or agouti-related peptide (AgRP) neurons stimulating feeding. Insulin and leptin exert catabolic effects on energy balance by decreasing food intake and stimulating metabolic rate, whereas ghrelin is anabolic and induces the opposite effects in mammals (Wren et al., 2000).

Insulin, the critical regulator of plasma glucose, exerts many of its physiological effects on body weight regulation by acting on target neurons in the IN (Shiraishi et al., 2011a, b). NPY is one of the downstream mediators of insulin action in the hypothalamus, which is supported by the observation that insulin receptors are present in NPY neurons in the IN (Shiraishi et al., 2011a). Additionally, central injection of insulin inhibits food intake with decreased gene expression of NPY in the hypothalamus (Shiraishi et al., 2008). Leptin in the hypothalamic ARC facilitates satiety through inhibition of the NPY neurons in mammals (Sahu, 2003). In chickens, the effect of central administration of leptin is consistent with an inhibitory influence of the hormone on feed intake (Denbow et al., 2000; Dridi et al., 2000), and continuous infusion of chicken leptin down-regulates the NPY receptor mRNA expression in the hypothalamus (Dridi et al., 2005). Ghrelin was shown to stimulate feeding behavior (Strader and Woods, 2005) by its action on NPY neurons in the central nervous system (Cowley et al., 2003; Greenman et al., 2004), suggesting the role of the melanocortin pathway in transmitting gut hunger signals to the mammalian brain. In chickens, dissimilar to mammalian results, central administration of ghrelin has the opposite effect on feeding: it prevents NPY-induced hyperphagia (Saito et al., 2002; 2005). The level of NPY mRNA in the central nervous system is not altered by ghrelin injection (Saito et al., 2005).

Figure 2 shows the proposed model of melanocortin system in mammals: interactions between NPY/AgRP and POMC neurons, and autoregulatory feedback from opioid and melanocortin peptides as well as NPY (e.g., Cowley et al., 2001; Jobst et al., 2004). In this model, both POMC and NPY neurons express autoreceptors for some of their respective neuropeptide products (β-endorphin or α-melanocyte stimulating hormone (MSH), and NPY, respectively) and activation of these autoreceptors may provide ultrashort feedback loops that further modulate the effects of peripheral signals (e.g., insulin) on POMC neurons.

With regard to the relationship with POMC-related agents on NPY-induced hyperphagia in chickens, co-injection of α-MSH or μ-opioid receptor antagonists are effective in reducing NPY-induced feeding in chicks (Kawakami et al., 2000; Cline and Smith, 2007; Dodo et al., 2005). From the above model, reduction of NPY-induced feeding by α-MSH (Kawakami et al., 2000; Cline and Smith, 2007) might be attributed to attenuation of NPY neuron activity via MC3-R. In addition, α-MSH also acts on the down-stream neurons, such as corticotropin-releasing hormone, eliciting corticosterone release (Tachibana et al., 2007). The attenuated effect of β-endorphin via μ-opioid receptors on POMC neurons could be blocked by antagonists. Similar to α-MSH, other anorexic peptides also suppress the NPY-induced feeding in chickens: glucagon-like peptide-1 completely inhibited the effect of NPY on food intake without a dose response (Furuse et al., 1997), and co-injection of bombesin attenuated the hyperphagia by NPY in a dose-related fashion (Bungo et al., 2000). Such findings suggest that these peptides are the second-order messengers of the melanocortin system. Anyway, feeding behavior in chickens is regulated by anorexic agents but not orexigenic ones, such as NPY (Furuse et al., 2007).

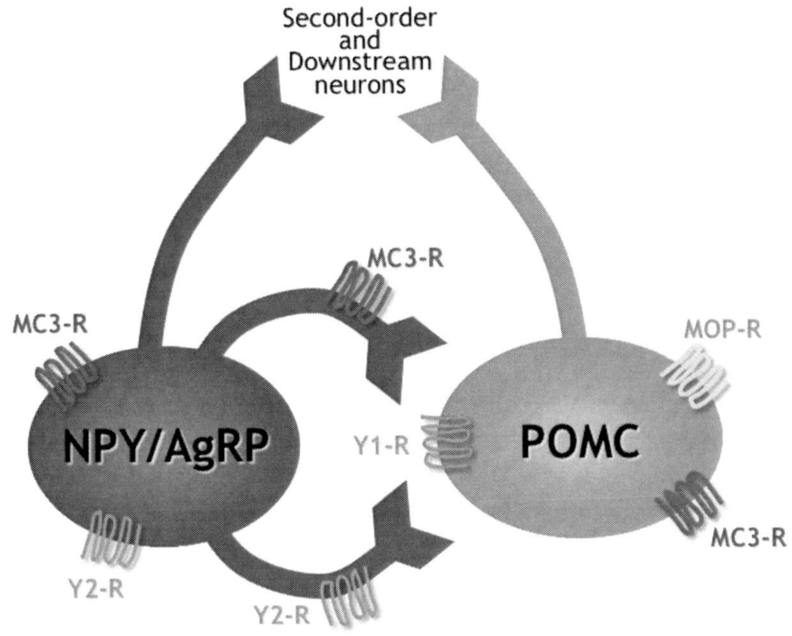

Figure 2. The proposal model of melanocortin system in the mammalian ARC. NPY: neuropeptide Y, AgRP: agouti-related Peptide, POMC: pro-opiomelanocortin, Y1-R: Y1 receptor, Y2-R: Y2 receptor, MC3-R: melanocortin-3 receptor, MOP-R: mu-opioid receptor.

Although as mentioned above, NPY Y1 and Y5 receptors might be "feeding receptors" for the action of NPY on food intake in chickens, pharmacological studies have shown contradictory results: the Y1 receptor antagonist did not attenuate NPY-induced feeding (Kawakami et al., 2001), but [Leu31, Pro34]-NPY, an agonist with appreciable affinity for the chicken NPY Y2 and Y4 receptors (Salaneck et al., 2000; Lundell et al., 2002), increased feed intake comparable to NPY (Ando et al., 2001). Moreover, our recent study showed that oligodeoxynucleotides against NPY Y1 receptor suppressed feed intake in chicks fed *ad libitum* (Bungo et al., unpublished data). The Y1 and Y2 receptors seem to have acquired antagonistic roles in the mammal due to a triple mutation in the Y2 receptor (Berglund et al., 2002), but could even functionally complement in the chick. Further studies will be needed to understand the role of NPY and its receptor subtypes in the melanocortin system of the chick.

Age and Strain Differences

Layer and broiler chickens have been intensively selected over generations for egg and meat production respectively and demonstrate striking differences in body weight and body composition. The difference could be due partly to the difference in food intake between two types of chickens (Masic et al., 1974; Savory, 1974; Denbow, 1999). There is evidence showing a significant increase in food intake in broiler chickens compared to layer chickens from the neonatal period (Nir et al., 1993; Mahagna and Nir, 1996). In addition, NPY is known as one of the most potent endogenous stimulators of feeding in chickens (Kuenzel et al., 1987; Furuse et al., 1997). Numerous research efforts have focused on the hypothalamic NPY as a promising approach for the regulatory mechanisms of feeding and metabolism in chickens.

Unexpectedly, hypothalamic NPY seems not to be the main agent promoting hyperphagia in broiler chickens. The level of NPY mRNA was significantly affected by age, but not by genotype at 1, 9, and 35 days of age (Cassy et al., 2004). Yuan and co-workers (2009) also reported that broiler and layer chicks exhibit similar hypothalamic expression of NPY at 7 days of age. While no breed differences on the expression of the NPY genes were found, a higher expression of NPY mRNA was detected in the PVN, IN and LHy of layer chickens compared with that of broiler chickens at the age of 14 days (Chen et al., 2007). These results agree with other studies in which the hypothalamic NPY peptide content was greater in the layer-type than in the meat-type chicken (in the embryonic stage, Zhou et al., 2006; at 1, 2, and 4 days of age, Saneyasu et al., 2011). Similarly, hypothalamic NPY gene expression was higher in lean line chickens compared to fat line chickens at 9 weeks of age under *ad libitum* feeding, but that difference was not present in fasted state (Dridi et al., 2006). Comparison of the gene expression of hypothalamic NPY receptors of broilers and layers showed similar or lower levels in broiler than in layer chicks, the mRNA level of the Y1 at 2 and 4 days of age and Y5 receptors at 4 days of age were lower in broiler chicks than in layer chicks (Saneyasu et al., 2011). In any case, these findings suggest that the hyperphagia in broiler chicks is not caused by the increased level of the hypothalamic NPY mRNA or the density of its receptors.

From the studies of Denbow (1985) and Kuenzel (1989) it is clear that not only the hypothalamus, which is directly involved in the control of food intake, but also the forebrain structures such as the neostriatum, the hyperstriatum and the paleostriatal complex could be involved in the control of feeding in chickens. Cassy et al. (2004) suggested that the hyperphagic feeding of broiler chickens might result from lower sensitivity to anorectic factors (such as leptin), and not from higher levels of orexigenic factors than found in layer chickens. Other factors have been demonstrated to exert differential effects in layer and broiler chickens. For example, 5-hydroxytryptamine inhibits food intake in layer chicks, whereas it has no effect on broiler chickens (Denbow et al., 1982, 1983).

Similarly, anorexic effect of endogenous glucagon-like peptide 1 seems to be lower in broiler chicks compared to layer chicks (Tachibana et al., 2001). Growth hormone inhibits the hypothalamic NPY in chickens (Wang et al., 2000), and the peak of growth hormone secretion appears at 14-28 days of age (Burnside and Cogburn, 1992). Body weight of the meat-type chickens was significantly greater than that of the layer-type chickens at the same age, indicating a more rapid growing rate in the broiler chickens.

Conclusion

Although there are some differences, the amino acid sequences and/or the neuroanatomical distribution of NPY and its receptors are conserved in chickens and mammals (Larhammar, 1996; Salaneck et al., 2000; Berglund et al., 2002; Lundell et al., 2002). Also, the functions of NPY in the brain are likely to have been conserved during vertebrate evolution. This supports the role for NPY as a critical and central component of feeding regulation in chickens. There are, however, a number of unresolved questions regarding NPY and appetite.

Understanding the direct and indirect dynamics of NPY action can enable the control of appetite in chickens. Additionally, the knowledge gained from studying NPY regulation could be applied to other neuropeptide and neurotransmitter systems in order to develop a detailed understanding of the regulation of feeding behavior and energy homeostasis in chickens as well as in mammals.

Acknowledgments

The authors are grateful to all colleagues of the laboratory of Animal Behavior and Physiology, Faculty of Applied Biological Science, Hiroshima University, for taking care of the birds.

This work was partially supported by a Grant-in-Aid for Scientific Research from Japan Society for the Promotion of Science.

References

Aakerlund, L., Gether, U., Fuhlendorff, J., Schwartz, T.W. and Thastrup, O. (1990) Y1 receptors for neuropeptide Y are coupled to mobilization of intracellular calcium and inhibition of adenylate cyclase. *FEBS Letters*, 260: 73-78.

Allen, Y.S., Adrian, T.E., Allen, J.M., Tatemoto, K., Crow, T.J., Bloom, S.R. and Polak, J.M. (1983) Neuropeptide Y distribution in the rat brain. *Science*, 221: 877-879.

Ando, R., Kawakami, S.I., Bungo ,T., Ohgushi, A., Takagi, T., Denbow, D.M. and Furuse, M. (2001) Feeding responses to several neuropeptide Y receptor agonists in the neonatal chick. *European Journal of Pharmacology*, 427: 53-59.

Bard, J.A., Walker, M.W., Branchek, T.A. and Weinshank, R.L. (1995) Cloning and functional expression of a human Y4 subtype receptor for pancreatic polypeptide, neuropeptide Y, and peptide YY. *Journal of Biological Chemistry*, 270: 26762-26765.

Benoit, S.C., Air, E.L., Coolen, L.M., Strauss, R., Jackman, A., Clegg, D.J., Seeley, R.J. and Woods, S.C. (2002) The catabolic action of insulin in the brain is mediated by melanocortins. *Journal of Neuroscience*, 22: 9048-9052.

Berglund, M.M., Fredriksson, R., Salaneck, E. and Larhammar, D. (2002) Reciprocal mutations of neuropeptide Y receptor Y2 in human and chicken identify amino acids important for antagonist binding. *FEBS Letters*, 518: 5-9.

Billington, C.J., Briggs, J.E., Harker, S., Grace, M. and Levine, A.S. (1994) Neuropeptide Y in hypothalamic paraventricular nucleus: a center coordinating energy metabolism. *American Journal of Physiology, Regulatory Integrative Comparative Physiology*, 266: R1765-R1770.

Boswell, T., Millam, J.R., Li, Q. and Dunn, I.C. (1998) Cellular localization of neuropeptide Y mRNA and peptide in the brain of the Japanese quail and domestic chicken. *Cell and Tissue Research*, 193: 31-38.

Boswell, T., Dunn, I.C. and Corr, S.A. (1999a). Neuropeptide Y gene expression in the brain is stimulated by fasting and food restriction in chickens. *British Poultry Science,* 40: S42–S61.

Boswell, T., Dunn, I.C. and Corr, S.A. (1999b) Hypothalamic neuropeptide Y mRNA is increased after food restriction in growing broilers. *Poultry Science*, 78: 1203-1207.

Boswell, T., Li, Q. and Takeuchi, S. (2002) Neurons expressing neuropeptide Y mRNA in the infundibular hypothalamus of Japanese quail are activated by fasting and co-express agouti-related protein mRNA. *Molecular Brain Research*, 100: 31-42.

Bromée, T., Sjödin, P., Fredriksson, R., Boswell, T., Larsson, T.A., Salaneck, E., Zoorob, R., Mohel, N. and Larhammar, D. (2006) Neuropeptide Y-family receptors Y_6 and Y_7 in chicken. Cloning, pharmacological characterization, tissue distribution and conserved synteny with human chromosome region. *FEBS Journal*, 273: 2048-2063.

Bungo, T., Ando, R., Kawakami, S.I., Ohgushi, A., Shimojo, M., Masuda, Y. and Furuse, M. (2000) Central bombesin inhibits food intake and the orexigenic effect of neuropeptide Y in the neonatal chick. *Physiology and Behavior*, 70: 573-576.

Burkhoff, A.M., Linemeyer, D.L. and Salon, J.A. (1998) Distribution of a novel hypothalamic neuropeptide Y receptor gene and its absence in rat. *Molecular Brain Research*, 53: 311-316.

Burnside, J. and Cogburn, L.A. (1992) Developmental expression of hepatic growth hormone receptor and insulin-like growth factor-1 mRNA in the chicken. *Molecular and Cellular Endocrinology*, 89: 91-96.

Cassy, S., Picard, M., Crochet, S., Derouet, M., Keisler, D.H. and Taouis, M. (2004) Peripheral leptin effect on food intake in young chickens is influenced by age and strain. *Domestic Animal Endocrinology*, 27: 51-61.

Chance, W.T., Sheroff, S., Foley-Nelson, T., Fischer, J.E. and Balasubramaniam, A. (1989) Pertussis toxin inhibits neuropeptide Y-induced feeding in rats. *Peptides*, 10: 1283-1286.

Chen, G.Q., Hu, X.F., Sugahara, K., Chen, J.S., Song, X.M., Zheng, H.C., Jiang, Y.Q., Huang, X., Jiang, J.F. and Zhou, W.D. (2007) Type-dependent differential expression of neuropeptide Y in chicken hypothalamus (*Gallus domesticus*). *Journal of Zhejiang University Science B*, 8: 839–844.

Clark, J.T., Kalra, P.S., Crowley, W.R. and Kalra, S.P. (1984) Neuropeptide Y and human pancreatic polypeptide stimulates feeding behavior in rats. *Endocrinology*, 115: 427-429.

Cline, M.A. and Smith, M.L. (2007) Central α-melanocyte stimulating hormone attenuates behavioral effects of neuropeptide Y in chicks. *Physiology and Behavior*, 91: 588-592.

Cowley, M.A., Smart, J.L., Rubinstein, M., Cerdán, M.G., Diano, S., Horvath, T.L., Cone, R.D. and Low, M.J. (2001) Leptin activates anorexigenic POMC neurons through a neural network in the arcuate nucleus. *Nature*, 411: 480-484.

Cowley, M.A., Cone, R.D., Enriori, P., Louiselle, I., Williams, S.M. and Evans, A.E. (2003) Electrophysiological actions of peripheral hormones on melanocortin neurons. *Annals of the New York Academy of Sciences*, 994: 175-186.

Davies, L., and Marks, J.L. (1994) Role of hypothalamic neuropeptide Y gene expression in body weight regulation. *American Journal of Physiology, Regulatory Integrative Comparative Physiology*, 266: R1687-R1691.

Denbow, D.M., Van Krey, H.P., Lacy, M.P. and Dietrick, T.J. (1983) Feeding, drinking and body temperature of Leghorn chicks: effects of ICV injections of biogenic amines. *Physiology and Behavior*, 31: 85-90.

Denbow, D.M., Van Krey, H.P. and Cherry, J.A. (1982) Feeding and drinking response of young chicks to injections of serotonin into the lateral ventricle of the brain. *Poultry Science*, 61: 150-155.

Denbow, D.M. (1999) Food intake regulation in birds. *Journal of Experimental Zoology*, 283: 333-338.

Denbow, D.M. (1985) Food intake control in birds. *Neuroscience and Biobehavioral Reviews*, 9: 223-232.

Denbow, D.M., Meade, S., Robertson, A., McMurtry, J.P., Richards, M. and Ashwell, C. (2000) Leptin-induced decrease in food intake in chickens. *Physiology and Behavior*, 69: 359-362.

Dodo, K.-I., Izumi, T., Ueda, H. and Bungo, T. (2005) Response of neuropeptide Y-induced feeding to μ-, δ- and κ-opioid receptor antagonists in the neonatal chick *Neuroscience Letters*, 373: 85-88.

Dridi, S., Ververken, C., Hillgartner, F.B., Arckens, L., Van der, G.E., Cnops, L., Decuypere, E. and Buyse, J. (2006) FAS inhibitor cerulenin reduces food intake and melanocortin receptor gene expression without modulating the other (an)orexigenic neuropeptides in chickens. *American Journal of Physiology, Regulatory Integrative Comparative Physiology*, 291: R138-R147.

Dridi, S., Raver, N., Gussakovsky, E.E., Derouet, M., Picard, M., Gertler, A. and Taouis, M. (2000) Biological activities of recombinant chicken leptin C4S analog compared with unmodified leptons. *American Journal of Physiology*, 279: E116-E123.

Dridi, S., Dwennen, Q., Decuypere, E. and Buyse, J. (2005) Mode of leptin action in chicken hypothalamus. *Brain Research*, 1047: 214-223.

Dube, M.G., Sahu, A., Kalra, P.S. and Kalra, S.P. (1992) Neuropeptide Y release is elevated from the microdissected paraventricular nucleus of food-deprived rats: an in vitro study. *Endocrinology*, 131: 684-688.

Esposito, V., Pelagalli, G.V., De Girolamo, P. and Gargiulo, G. (2001) Anatomical distribution of NPY-like immunoreactivity in the domestic chick brain (*Gallus domesticus*). *Anatomical Record*, 263: 186–201.

Fredriksson, R., Larson, E.T., Yan, Y.-L., Postlethwait, J.H. and Larhammar, D. (2004) Novel neuropeptide Y Y2-like receptor subtype in zebrafish and frogs supports early vertebrate chromosome duplications. *Journal of Molecular Evolution*, 58: 106-114.

Freeman, B.M. (1965) The relationship between oxygen consumption, body temperature and surface area in the hatching and young chick. *British Poultry Science*, 6: 67-72.

Furuse, M., Matsumoto, M., Mori, R., Sugahara, K., Kano, K. and Hasegawa, S. (1997) Influence of fasting and neuropeptide Y on the suppressive food intake induced by intracerebroventricular injection of glucagon-like peptide-1 in the neonatal chick. *Brain Research*, 764: 289-292.

Furuse, M., Yamane, H., Tomonaga, S., Tsuneyoshi, Y. and Denbow, D.M. (2007) Neuropeptidergic regulation of food intake in the neonatal chick: A Review. *Journal of Poultry Science*, 44: 349-356.

Gerald, C.M., Walker, W., Vaysse, P.J., He, C., Branchek, T.A. and Weinshank, R.L. (1995) Expression cloning and pharmacological characterization of a human hippocampal neuropeptide Y/peptide YY Y2 receptor subtype. *Journal of Biological Chemistry*, 270: 26758-26761.

Greenman, Y., Golani, N., Gilad, S., Yaron, M., Limor, R. and Stren, N. (2004) Ghrelin secretion is modulated in a nutrient- and gender-specific manner. *Clinical Endocrinology*, 60: 382-386.

Gregor, P., Millham, M.L., Feng, Y., DeCarr, L.B., McCaleb, M.L. and Cornfield, L.J. (1996) Cloning and characterization of a novel receptor to pancreatic polypeptide, a member of the neuropeptide Y receptor family. *FEBS Letters*, 381: 58-62.

Herzog, H., Hort, Y.J., Shine, J. and Selbie, L.A. (1993) Molecular cloning, characterization and localization of the human homolog to the reported bovine NPY Y3 receptor: Lack of NPY binding and activation. *DNA and Cell Biology*, 12: 465-471.

Herzog, H., Darby, K., Ball, H., Hort, Y., Beck-Sickinger, A. and Shine, J. (1997) Overlapping gene structure of the human neuropeptide Y receptor subtypes Y1 and Y5 suggests coordinate transcriptional regulation. *Genomics*, 41: 315-319.

Higgins, S.E., Ellestad, L.E., Trakooljul, N., McCarthy, F., Saliba, J., Cogburn, L.A. and Porter, T.E. (2010) Transcriptional and pathway analysis in the hypothalamus of newly hatched chicks during fasting and delayed feeding. *BMC Genomics*, 11:162.

Holmberg, S.K., Mikko, S., Boswell, T., Zoorob, R. and Larhammar, D. (2002) Pharmacological characterization of cloned chicken neuropeptide Y receptors Y1 and Y5. *Journal of Neurochemistry*, 81: 462-471.

Horvath, T.L., Diano, S., Sotonyi, P., Heiman, M. and Tschöp, M. (2001) Minireview: ghrelin and the regulation of energy balance-a hypothalamic perspective. *Endocrinology*, 142: 4163-4169.

Jazin, E.E., Yoo, H., Blomqvist, A.G., Yee, F., Weng, G., Walker, M.W., Salon, J., Larhammar, D. and Wahlestedt, C. (1993) A proposed bovine neuropeptide Y (NPY) receptor, or its human homologue, confers neither NPY binding sites nor NPY responsiveness on transfected cells. *Regulatory Peptides*, 47: 247-258.

Jobst, E.E., Enriori, P.J. and Cowley, M.A. (2004) The electrophysiology of feeding circuits. *Trends in Endocrinology and Metabolism*, 15: 488-499.

Kalra, S.P., Dube, M.G., Sahu, A., Phelps, C.P. and Kalra, P.S. (1991) Neuropeptide Y secretion increases in the paraventricular nucleus in association with increased appetite for food. *Proceedings of the National Academy of Sciences, USA*, 88: 10931-10935.

Kawakami, S.I., Bungo, T., Ando, R., Ohgushi, A., Shimojo, M., Masuda, Y. and Furuse, M. (2000) Central administration of alpha-melanocyte stimulating hormone inhibits fasting- and neuropeptide Y-induced feeding in neonatal chicks. *European Journal of Pharmacology*, 398: 361-364.

Kawakami, S.I., Ando, R., Bungo, T., Ohgushi, A., Tachibana, T., Denbow, D.M. and Furuse, M. (2001) Receptor antagonist, does not inhibit fasting- and NPY-induced food intake in neonatal chicks. *Journal of Poultry Science*, 38: 259-265.

Kuenzel, W.J. (1989) Neuroanatomical substrates involved in the control of food intake. *Poultry Science*, 68: 926-937.

Kuenzel, W.J., Douglass, L.W. and Davison, B.A. (1987) Robust feeding following central administration of neuropeptide Y or peptide YY in chicks, *Gallus domesticus. Peptides*, 8: 823-828.

Kuenzel, W.J. and McMurtry, J. (1988) Neuropeptide Y: brain localization and central effects on plasma insulin levels in chicks. *Physiology and Behavior*, 44: 669-678.

Larhammar, D. (1996) Evolution of neuropeptide Y, peptide YY, and pancreatic polypeptide. *Regulatory Peptides*, 62: 1-11.

Leibowitz, S.F. (1989) Hypothalamic neuropeptide Y, galanin, and amines. Concepts of coexistence in relation to feeding behavior. *Annals of the New York Academy of Sciences*, 575: 221-233.

Lundell, I., Blomqvist, A.G., Berglund, M.M., Schober, D.A., Johnson, D., Statnick, M.A., Gadski, R.A., Gehlert, D.R. and Larhammar, D. (1995) Cloning of a human receptor of the NPY receptor family with high affinity for pancreatic polypeptide and peptide YY. *Journal of Biological Chemistry*, 270: 29123-29128.

Lundell, I., Statnick, M.A., Johnson, D., Schober, D.A., Starback, P., Gehlert, D.R. and Larhammar, D. (1996) The cloned rat pancreatic polypeptide receptor exhibits profound differences to the orthologous receptor. *Proceedings of the National Academy of Sciences, USA*, 93: 5111-5115.

Lundell, I., Boswell, T. and Larhammar, D. (2002) Chicken neuropeptide Y-family receptor Y4: a receptor with equal affinity for pancreatic polypeptide, neuropeptide Y and peptide YY. *Journal of Molecular Endocrinology*, 28: 225-235.

Mahagna, M. and Nir, I. (1996) Comparative development of digestive organs, intestinal disacchridases and some blood metabolites in broiler and layer-type chicks after hatching. *British Poultry Science*, 37: 359–371.

Masic, B., Wood-Gush, D.G.M., Duncan, I.J.H., McCorquodale, C. and Savory, C.J. (1974) A comparison of feeding behavior of young broiler and layer males. *British Poultry Science*, 15: 281-286.

Merckaert, J. and Vandesande, F. (1996) Autoradiographic localization of receptors for neuropeptide Y (NPY) in the brain of broiler and leghorn chickens (*Gallus domesticus*). *Journal of Chemical Neuroanatomy*, 12: 123-134.

Michel, M.C., Beck-Sickinger, A., Cox, H., Doods, H.N., Herzog, H., Larhammar, D., Quirion, R., Schwartz, T. and Westfall, T. (1998) XVI. International Union of Pharmacology recommendations for the nomenclature of neuropeptide Y, peptide YY and pancreatic polypeptide receptors. *Pharmacological Reviews*, 50: 143-150.

Morley, J.E., Levine, A.S., Gosnell, B.A., Kneip, J. and Grace, M. (1987) Effect of neuropeptide Y on ingestive behaviors in the rat. *American Journal of Physiology, Regulatory Integrative Comparative Physiology*, 252: R599-R609.

Nakamura, M., Yokoyama, M., Watanabe, H. and Matsumoto, T. (1997) Molecular cloning, organization and localization of the gene for the mouse neuropeptide Y-Y5 receptor. *Biochimical et Biophysica Acta*, 1328: 83-89.

Nir, I., Nitsan, Z. and Mahagna, M. (1993) Comparative growth and development of the digestive organs and of some enzymes in broiler and egg type chicks after hatching. *British Poultry Science*, 34: 523-532.

Rose, P.M., Fernandes, P., Lynch, J.S., Frazier, S.T., Fisher, S.M., Kodukula, K., Kienzle, B. and Seethala, R. (1995) Cloning and functional expression of a cDNA encoding a human type 2 neuropeptide Y receptor. *Journal of Biological Chemistry*, 270: 22661-22664.

Sahu, A., Kalra, S.P., Crowley, W.R. and Kalra, P.S. (1988) Evidence that NPY-containing neurons in the brain stem project into selected hypothalamic nuclei implication in feeding behavior. *Brain Research*, 457: 376-378.

Sahu, A., White, J.D., Kalra, P.S. and Kalra, S.P. (1992) Hypothalamic neuropeptide Y gene expression in rats on scheduled feeding regimen. *Molecular Brain Research*, 15: 15-18.

Sahu, A. (2003) Leptin signaling in the hypothalamus: emphasis on energy homeostasis and leptin resistance. *Frontiers in Neuroendocrinology*, 24: 225-253.

Saito, E.S., Kaiya, H., Takagi, T., Yamasaki, I., Denbow, D.M., Kangawa, K. and Furuse, M. (2002) Chicken ghrelin and growth hormone-releasing peptide-2 inhibit food intake of neonatal chicks. *European Journal of Pharmacology*, 453: 75-79.

Saito, E.S., Kaiya, H., Tachibana, T., Tomonaga, S., Denbow, D.M., Kangawa, K. and Furuse, M. (2005) Inhibitory effect of ghrelin on food intake is mediated by the corticotropin-releasing factor system in neonatal chicks. *Regulatory Peptides*, 125: 201-208.

Salaneck, E., Holmberg, S.K., Berglund, M.M., Boswell, T. and Larhammar, D. (2000) Chicken neuropeptide Y receptor Y2: structural and pharmacological differences to mammalian Y2. *FEBS Letters*, 484: 229-234.

Saneyasu, T., Honda, K., Kamisoyama, H., Ikura, A., Nakayama, Y. and Hasegawa, S. (2011) Neuropeptide Y effect on food intake in broiler and layer chicks. *Comparative Biochemistry and Physiology, Part A: Molecular and Integrative Physiology*, 159: 422-426.

Savory, C.J. (1974) A comparison of the feeding behaviour of young broiler and layer males. *British Poultry Science*, 15: 499-505.

Schwartz, M.W., Woods, S.C., Porte Jr., D., Seeley, R.J. and Baskin, D.G. (2000) Central nervous system control of food intake. *Nature*, 404: 661-671.

Shiraishi, J.-i., Yanagita, K., Fujita, M. and Bungo, T. (2008) Central insulin suppress feeding behavior via melanocortins in chicks. *Domestic Animal Endocrinology*, 34: 223-228.

Shiraishi, J.-i., Tanizawa, H., Fujita, M., Kawakami, S.-I. and Bungo, T. (2011a) Localization of hypothalamic insulin receptor in neonatal chicks: Evidence for insulinergic system control of feeding behavior. *Neuroscience Letters*, 491: 177-180.

Shiraishi, J.-i., Yanagita, K. Fukumori, R., Sugino, T., Fujita, M. Kawakami, S.-I., McMurtry, J.P. and Bungo, T. (2011b) Comparisons of insulin related parameters in commercial-type chicks: Evidence for insulin resistance in broiler chicks. *Physiology and Behavior*, 103: 233-239.

Starbäck, P., Wraith, A., Eriksson, H. and Larhammar, D. (2000) Neuropeptide Y receptor gene y6: multiple deaths or resurrections? *Biochemical and Biophysical Research Communications*, 277: 264-269.

Steinman, J.L., Denson, L., Fujikawa, G., Wasterlain, C.G., Cherkin, A. and Morley, J.E. (1987) The effects of adrenergic, opioid and pancreatic polypeptidergic compounds on feeding and other behaviors in neonatal leghorn chicks. *Peptides,* 8: 585-592.

Strader, A.D. and Woods, S.C. (2005) Gastrointestinal hormones and food intake. *Gastroenterology*, 128: 175-191.

Tachibana, T., Sugahara, K., Ohgushi, A., Ando, R., Sashihara, K., Yoshimatsu, T. and Furuse, M. (2001) Intracerebroventricular injection of exendin (5–39) increases food intake of layer-type chicks but not broiler chicks. *Brain Research*, 915: 234-237.

Tachibana, T., Saito, S., Tomonaga, S., Takagi, T., Saito, E.S., Nakanishi, T., Koutoku, T., Tsukada, A., Ohkubo, T., Boswell, T. and Furuse, M. (2004) Effect of central administration of prolactin-releasing peptide on feeding in chicks. *Physiology and Behavior*, 80: 713-719.

Tachibana, T., Sato, M., Oikawa, D., Takahashi, H., Boswell, T. and Furuse, M. (2006) Intracerebroventricular injection of neuropeptide Y modifies carbohydrate and lipid metabolism in chicks. *Regulatory Peptides*, 136: 1-8.

Tachibana, T., Oikawa, D., Takahashi, H., Boswell, T. and Furuse, M. (2007) The anorexic effect of alpha-melanocyte-stimulating hormone is mediated by corticotrophin-releasing factor in chicks. *Comparative Biochemistry and Physiology Part A: Molecular and Integrative Physiology*, 147: 173-178.

Tatemoto, K., Carlquist, M. and Mutt, V. (1982) Neuropeptide Y-a novel brain peptide with structural similarities to peptide YY and pancreatic polypeptide. *Nature*, 296: 659-660.

Wahlestedt, C., Regunathan, S. and Reis, D.J. (1992) Identification of cultured cells selectively expressing Y1-, Y2-, or Y3-type receptors for neuropeptide Y/peptide YY. *Life Sciences,* 50: PL7-PL12.

Wang, X., Day, J.R. and Vasilatos-Younken, R. (2001) The distribution of neuropeptide Y gene expression in the chicken brain. *Molecular and Cellular Endocrinology*, 174: 129-136.

Wren, A.M., Small, C.J., Ward, H.L., Murphy, K.G., Dakin, C.L., Taheri, S., Kennedy, A.R., Roberts, G.H., Morgan, D.G., Ghatei, M.A. and Bloom, S.R. (2000) The novel hypothalamic peptide ghrelin stimulates food intake and growth hormone secretion. *Endocrinology*, 141: 4325-4328.

Yuan, L., Ni, Y., Barth, S., Wang, Y., Grossmann, R. and Zhao, R. (2009) Layer and broiler chicks exhibit similar hypothalamic expression of orexigenic neuropeptides but distinct expression of genes related to energy homeostasis and obesity. *Brain Research*, 1273: 18-28.

Zhou, W., Murakami, M., Hasegawa, S., Yoshizawa, F. and Sugahara, K. (2005) Neuropeptide Y content in the hypothalamic paraventricular nucleus responds to fasting and refeeding in broiler chickens. *Comparative Biochemistry and Physiology-Part A: Molecular and Integrative Physiology*, 141: 146-152.

Zhou, W., Aoyama, M., Yoshizawa, F. and Sugahara, K. (2006) Developmental increases in hypothalamic neuropeptide Y content with the embryonic age of meat- and layer-type chicks. *Brain Research*, 1072: 26-29.

In: Neuropeptide Y
Editors: Steven L. Parker

ISBN: 978-1-62618-421-3
© 2013 Nova Science Publishers, Inc.

Chapter II

Neuropeptide Y (NPY) Relations to Obesity and Metabolic Syndrome

Flávia Campos Corgosinho and Ana Raimunda Dâmaso*
Flávia Campos Corgosinho and Ana Raimunda Dâmaso
Biosciences Department, Federal University of Sao Paulo-Paulista
Medicine School-UNIFESP-EPM, Sao Paulo, Brazil

Abstract

Obesity is a public health problem worldwide, affecting both developed societies and developing countries. The central nervous system has evolved a meticulously interconnected circuitry in order to maintain energy homeostasis and adequate nutritional state. Neuropeptide Y (NPY), a 36-amino-acid peptide, is one of the most abundant and widely distributed peptides in the neural matrix. Among physiological actions of NPY, potent orexigenic effects have possibly the largest impact. NPY and its receptors have been implicated in a variety of physiological effects such as regulation of food intake, energy balance, hormone release, cardiovascular disease, thermoregulation, stress response and anxiety. There are at least four receptors commonly expressed across the vertebrates that specifically respond to NPY and its analogs. These

* E-mail: flaviacorgosinho@hotmail.com

include the structurally quite similar Y1, Y4 and Y5 subtypes and the rather divergent Y2 receptor. This complex system is embedded in a densely redundant network which helps stable energy homeostasis. Previously we showed that NPY increases in a compensatory manner during weight loss in obese adolescents. This review focuses on effects of NPY in obesity and the metabolic syndrome. We also examine the neuroendocrine regulation of food intake.

Keywords: Neuropeptide Y; Energy balance; Obesity; Metabolic Syndrome; Exercise

Introduction

Obesity is a public health problem worldwide, affecting both developed and developing countries (Shamseddeen et al., 2011). Obesity increases the risk for developing chronic diseases, including type 2 diabetes, coronary heart disease, hypertension, osteoarthritis, and certain cancers, besides the social and psychological consequences (Caranti et al., 2007; Carnier et al., 2008; de Lima Sanches et al., 2010; Munhoz et al., 2012; Tock et al., 2006).

The pandemic proportions for obesity and its financial burden for governments make it a major medical problem worldwide (Shamseddeen et al., 2011). Obesity is mainly caused by a dysregulation of the energy balance, although the central nervous system has developed a meticulously inter-connected circuitry in order to maintain energy homeostasis and adequate nutritional state (Broberger, 2005).

Neuropeptide Y (NPY) was discovered by Tatemoto and colleagues in 1982 (Tatemoto et al., 1982). Nowadays it is well known that NPY has a potent orexigenic action, and its central effects include feeding behavior, playing a key role in the control of appetite, body weight and obesity (Kamiji and Inui, 2007).

Systemic effects of NPY include cardiac and vascular actions, such as modulation of heart rate, coronary blood flow and ventricular function, and vasoconstriction (Jacques and Abdel-Samad, 2007). Recently our group showed that NPY levels are increased in a compensatory manner during weight loss in obese adolescents, with adverse effects in individuals presenting the metabolic syndrome (Corgosinho et al., 2011). The present review focuses on the role and effects of NPY in obesity and development of the metabolic syndrome.

The Molecular Properties of NPY

Neuropeptide Y is a 36-peptide containing five tyrosine residues (hence its name), and a C-terminal RFamide-like dipeptide (Dyzma et al., 2010). NPY is a member of the Y peptide family also containing peptide YY(PYY; also a five-tyrosine 36-peptide) and PYY-derived pancreatic polypeptide (PP; a 36-peptide with four tyrosines in man) (Larhammar, 1996). Human NPY has about 67% sequence identity with human PYY, and also a considerable identity with human PP. Neuropeptide Y presents a tertiary structure which displays a hairpin-like conformation, playing a crucial role in its binding to its specific receptor.

The tertiary structure of NPY (Lerch et al., 2005; Zou et al., 2009) has a much weaker helical back-folding ("PP fold") than that of avian pancreatic polypeptide (Blundell et al., 1981). The non-helical residues 31–36 (ITRQRY in NPY, VTRQRY in PYY, LTRPRY in PP) are responsible for high-affinity binding of NPY and other Y peptides to all Y receptors (Fuhlendorff et al., 1990; Gehlert et al., 1996).

The lack of rigid PP fold in NPY enhances the non-specific, protean reactivity of NPY compared to PYY and PP (for a recent review see (Parker et al., 2011)), confers avidity for lipids of the bilayer (McLean et al., 1990), and could largely be responsible for the non-saturating traffic of NPY through the endothelial brain-blood barrier (BBB) (Kastin and Akerstrom, 1999), as opposed to an essentially saturating entry of PYY(1-36) (Hernandez et al., 1994; Nonaka et al., 2003).

NPY acts through heptahelical G-protein coupling receptors (GPCRs). Four functional Y peptide receptors are thus far cloned in the mammal, the Y1, Y2, Y4 and Y5 subtypes (Michel et al., 1998). The Y4 receptor is selective for pancreatic polypeptides, accepts PYY at a much lower affinity than PP, and poorly binds NPY. This can be traced especially to position 34 (Gln in NPY and PYY, Pro in PP), and partly also to position 31 (Leu in PP and PYY, Ileu in NPY) (Gehlert et al., 1996; Gehlert et al., 1997). The presumed NPY-selective y3 receptor (Wahlestedt et al., 1992) thus far was not cloned. All Y receptors transduce via $G_{i/o}$ type α subunits to inhibit adenylyl cyclase activity (Holliday et al., 2004). Transduction to G_q subunit is known for the Y2 receptor (Grouzmann et al., 2001).

There is coupling of the Y1 receptor to G_q-type α subunit(s), demonstrable after knockout of $G_{i/o}$ by pertussis toxin (Parker et al., 2008). As is typical for family A rhodopsin-like receptors, the Y receptors are produced and stored as heteropentameric complexes of receptor dimers with G-protein heterotrimers

(Baneres and Parello, 2003). Table 1 lists the brain distribution of NPY receptors, with references about their main physiological activities.

Table 1. NPY receptors: the effects on regulation of feeding and localization

NPY receptor	Effects on feeding	Main localization	References
Y1	Stimulates food ingestion and appetitive behaviour	ARC VMN, PVN	(Chee et al., 2010) (Jacques et al., 1997)
Y2	Inhibits NPY release. Anorexigenic and angiogenic.	ARC, PVN	(Chee et al., 2010)
Y3	Not cloned (only pharmacological experiments)	Adrenal ?	(Jacques et al., 1997) (Bromee et al., 2006)
Y4	Inhibition of food intake	LHA	(Chee et al., 2010)
Y5	Related to consummatory behaviour and reduction of energy expenditure	ARC, PVN	(Kamiji and Inui, 2007) (Chee et al., 2010)

*PVN: paraventricular nucleus, ARC: arcuate nucleus; LHA: lateral hypothalamic area; PFA: perifornical area.

NPY and Neuroendocrine Regulation of Food Intake

Since the discovery of NPY many studies were performed to better understand its metabolic, behavioral, and mechanistic effects on energy homeostasis. Neuropeptide Y is the most abundant and potent orexigenic peptide found in mammalian brain (Adrian et al., 1983; Allen et al., 1983), and its main source within the hypothalamus are neurons in the arcuate nucleus (ARC) (Chronwall et al., 1985). NPY release precedes the onset of feeding and gradually decreases with food intake. Increases in NPY expression precede hyperphagia and are directly correlated with conditions of decreased energy expenditure. Considering that ARC neurons in turn send projections to various hypothalamic neurons, the actions of NPY extend throughout the hypothalamus and affect a diverse and heterogeneous population of neurons (Mercer et al., 2011).

NPY mediates its orexigenic activity in the hypothalamus suppressing the primary anorexigenic system in the ARC. Release of NPY from NPY/AgRP neurons in the arcuate nucleus activates postsynaptic Y1 receptors to directly inhibit anorexigenic arcuate nucleus POMC/CART neurons (possibly via Gi-linked decrease in cAMP output), and activates presynaptic receptors (which can include Y2 receptors acting via Gq α (Grouzmann et al., 2001)) to inhibit melanocortin release. Peripheral and central signals of energy balance are integrated within the ARC both via the NPY and melanocortin / MSH systems (Acuna-Goycolea and van den Pol, 2005; Batterham et al., 2002; Ghamari-Langroudi et al., 2005).

In the PVN, NPY modulates neuronal activity by both presynaptic and postsynaptic mechanisms. NPY inhibits GABA release from neurons within the PVN (Melnick et al., 2007).

The postsynaptic effect of NPY is evident specifically at PVN neurons that express the MC4R, an anorexigenic melanocortin receptor. The presynaptic effect of NPY is mediated by Y1, Y2, and Y5 subtypes. Interestingly, the postsynaptic response appears to be mediated entirely by Y1 receptors (Pronchuk et al., 2002).

NPY also acts in presynaptic and postsynaptic in the VMN, by inhibiting synaptic transmission and reducing their excitability respectively (Chance et al., 1991; Chee et al., 2010). NPY within the VMN is less abundant than in other hypothalamic regions, but its release in the VMN has a high significance. NPY suppresses the VMN-mediated excitation of ARC POMC neurons. This would join the postsynaptic inhibition by local release of NPY in the ARC, concomitantly suppressing the anorexigenic ARC cells (Chee et al., 2010).

Two major classes of feeding-affecting neurons are present in the LHA, the orexin and the melanin concentrating hormone (MCH) neurons, both with orexigenic action (Elias et al., 1998).

Interestingly, NPY does not excite the main orexigenic systems in the LHA. It is suggested that the LHA orexigenic systems act partly by helping to activate the ARC NPY system, and once activated, the NPY neurons in turn suppress the LHA neurons (Mercer et al., 2011). The proposed mechanisms in activation of appetite by NPY are illustrated in Figure 1.

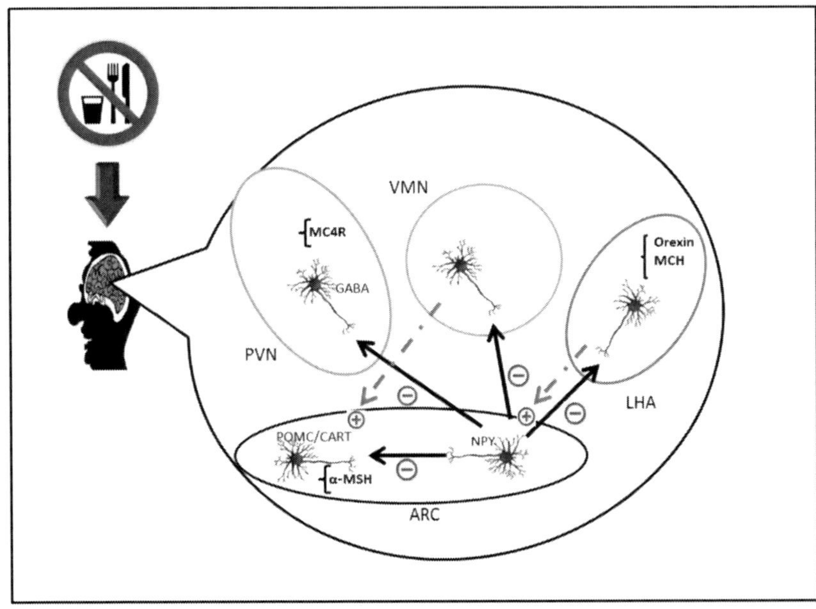

Figure 1. Main projections from NPY neurons in the arcuate nucleus (ARC), paraventricular nucleus (PVN), lateral hypothalamic area (LHA) and ventromedial nucleus (VMN). Within ARC, NPY inhibits anorexigenic neurons supressing α-MSH release. In the PVN the effect is an inhibition of GABA release from a population of GABAergic terminals, supressing the anorexigenic receptor MC4R. VMN activates POMC/CART in the ARC, however this is suppressed by NPY. LHA releases orexin and MCH, which activates NPY in the ARC. However NPY inhibits the release of orexins (Fu et al., 2004).

NPY and Obesity

As mentioned above, NPY is a potent orexigenic peptide found in the hypothalamus, stimulating food intake with a preferential effects on carbohydrate intake. NPY plays an important role in decreasing latency to eat, increases motivation to eat and delays satiety thus augmenting meal size. This results in increased body weight and feeding-induced obesity, and their co-morbidities (Beck, 2006).

In genetic models of obesity the identification of leptin ând its receptor, in the mid-1990s, was a pivotal point (Zhang et al., 1994). It was found that leptin signaling is low in Zucher fa/fa rat, the db/db mouse and the ob/ob mouse. The central NPY system is profoundly altered especially in the obese Zucher rat. Before discovery of leptin, this dysregulation was believed to be the primary cause in the development of obesity (Bray, 1993; Robinson, 1996).

Early increases in NPY content and expression in Zucher fatty rat have physiological consequences for NPY receptor function, contributing to upregulation of NPY system mediated by a deficiency in leptin signalling, which produces hyperphagia, excessive weight gain and obesity(Tritos et al., 1998). In fact, recently our group showed that in a state of hyperleptinemia obese adolescents have an increased NPY concentration during weight loss, suggesting an impairment in the control of energy balance and obesity. The vicious cycle of hyperleptinemia and upregulation of NPY system may however be broken by improvements in food ingestion and exercise as non-pharmacological strategies (Damaso et al., 2011).

Moreover, NPY was described to present a negative correlation with basal metabolic rate in obese adolescents. This reinforces the important role of NPY in energy balance (Carnier et al., 2008).

NPY and Metabolic Syndrome

The role of NPY in metabolic syndrome control is not clear. The metabolic syndrome (MetS) is characterized by a constellation of clinical and biochemical features that increase the risk of cardiovascular disease (Caranti et al., 2007). NPY enhances the development of obesity and other aspects of MetS. Recently, association between NPY Leu7Pro polymorphism and features of MetS have been studied in Iranian patients with coronary artery disease (CAD).

A significantly higher frequency of the Leu7Pro polymorphism was found in patients with MetS. Furthermore, there was a significant difference in Pro7 frequency between diabetics versus non-diabetics, dyslipidemic versus non-dyslipidemic, and obese versus non-obese in this population. Leu7Pro polymorphism could be associated with the MetS in patients with CAD (Masoudi Kazemabad et al., 2012).

Data from our research show that obese adolescents with non-alcoholic fatty liver disease (NAFLD) have an inverse correlation between NPY and adiponectin levels.

These results indicate that an altered pattern of the NPY system may promote changes in energy balance associated with inflammatory process in obesity and its co-morbidities. NAFLD has indicated a new biomarker of the metabolic syndrome, a complex metabolic altered milieu inducing cardiovascular diseases (de Piano et al., 2011). Moreover, it was shown that the increase in visceral fats is an independent predictor of increases in NPY, promoting a positive energy balance in obese adolescents with NAFLD (de Piano et al., 2010). Indeed, a recent study from our group, comparing adolescents with and without metabolic syndrome showed that those with metabolic syndrome present, after weight loss, significant increase in NPY levels, which could hamper the maintenance of the new weight (Corgosinho et al., 2011).

NPY and Exercise

In a long term therapy, the increased level of NPY in the first step of weight loss may suggest a compensatory response in resetting the energy balance. However, when the obese people were maintained on a weight loss therapy and aerobic *plus* resistance training as combined tools, after a reduction of approximately 10% in body weight both hyperleptinemia and NPY were improved, promoting a negative energy balance. These adaptations induce increase in α-MSH, an important anorexigenic neuropeptide (Oyama et al., 2010).

Moreover, the multidisciplinary therapy using nutritional counseling and aerobic exercise decreased NPY and AgRP values and increased α-MSH. Simultaneous with these changes there was a decrease in total food intake after 24 weeks of therapy (Prado et al., 2011). However, it is necessary to further explore what kind of exercise would be the most effective in the control of NPY (with and without hyperleptinemia).

Conclusion and Future Directions

The complex set of genetic, physiological and behavioral findings considered in this review involves several hypothalamic neuropeptides as the key regulators of food intake. Rather than simply increasing (or decreasing as the case may be) food intake, it is now apparent that hypothalamic peptides might play significant roles in more complex behavioral ând psychological processes.

Specifically, these signals may play important roles in the generation of state cues (e.g., hunger), macro-nutrient selection, and cognitive, inflammatory and weight loss processes. We suggest that a more comprehensive and inclusive understanding of these intake processes may help to develop novel therapeutic interventions for the treatment of obesity and its co-morbidities, especially as related to behavioral disorders.

References

Acuna-Goycolea C and van den Pol AN (2005) Peptide YY(3-36) inhibits both anorexigenic proopiomelanocortin and orexigenic neuropeptide Y neurons: implications for hypothalamic regulation of energy homeostasis. *J. Neurosci.* 25(45):10510-10519.

Adrian TE, Allen JM, Bloom SR, Ghatei MA, Rossor MN, Roberts GW, Crow TJ, Tatemoto K and Polak JM (1983) Neuropeptide Y distribution in human brain. *Nature* 306(5943):584-586.

Allen YS, Adrian TE, Allen JM, Tatemoto K, Crow TJ, Bloom SR and Polak JM (1983) Neuropeptide Y distribution in the rat brain. *Science* 221(4613):877-879.

Baneres JL and Parello J (2003) Structure-based analysis of GPCR function: evidence for a novel pentameric assembly between the dimeric leukotriene B4 receptor BLT1 and the G-protein. *J. Mol. Biol.* 329(4):815-829.

Batterham RL, Cowley MA, Small CJ, Herzog H, Cohen MA, Dakin CL, Wren AM, Brynes AE, Low MJ, Ghatei MA, Cone RD and Bloom SR (2002) Gut hormone PYY(3-36) physiologically inhibits food intake. *Nature* 418(6898):650-654.

Beck B (2006) Neuropeptide Y in normal eating and in genetic and dietary-induced obesity. *Philos Trans R Soc Lond B Biol Sci* 361(1471):1159-1185.

Blundell TL, Pitts JE, Tickle IJ, Wood SP and Wu CW (1981) X-ray analysis (1. 4-A resolution) of avian pancreatic polypeptide: Small globular protein hormone. *Proc. Natl. Acad. Sci. U S A* 78(7):4175-4179.

Bray GA (1993) The nutrient balance hypothesis: peptides, sympathetic activity, and food intake. *Ann. N. Y. Acad. Sci.* 676:223-241.

Broberger C (2005) Brain regulation of food intake and appetite: molecules and networks. *J. Intern. Med.* 258(4):301-327.

Bromee T, Sjodin P, Fredriksson R, Boswell T, Larsson TA, Salaneck E, Zoorob R, Mohell N and Larhammar D (2006) Neuropeptide Y-family receptors Y6 and Y7 in chicken. Cloning, pharmacological characterization, tissue distribution and conserved synteny with human chromosome region. *Febs. J.* 273(9):2048-2063.

Caranti DA, Tock L, Prado WL, Siqueira KO, de Piano A, Lofrano M, Damaso AR, Damaso AR, Cristofalo DM, Lederman H, Damaso AR, de Mello MT, Tufik S, Damaso AR and Damaso AR (2007) Long-term multidisciplinary therapy decreases predictors and prevalence of metabolic syndrome in obese adolescents. *Nutr. Metab. Cardiovasc. Dis.* 17(6):e11-13.

Carnier J, Lofrano MC, Prado WL, Caranti DA, de Piano A, Tock L, do Nascimento CM, Oyama LM, Mello MT, Tufik S and Damaso AR (2008) Hormonal alteration in obese adolescents with eating disorder: effects of multidisciplinary therapy. *Horm. Res.* 70(2):79-84.

Chance WT, Balasubramaniam A, Zhang FS, Wimalawansa SJ and Fischer JE (1991) Anorexia following the intrahypothalamic administration of amylin. *Brain Res.* 539(2):352-354.

Chee MJ, Myers MG, Jr., Price CJ and Colmers WF (2010) Neuropeptide Y suppresses anorexigenic output from the ventromedial nucleus of the hypothalamus. *J. Neurosci.* 30(9):3380-3390.

Chronwall BM, DiMaggio DA, Massari VJ, Pickel VM, Ruggiero DA and O'Donohue TL (1985) The anatomy of neuropeptide-Y-containing neurons in rat brain. *Neuroscience* 15(4):1159-1181.

Corgosinho FC, de Piano A, Sanches PL, Campos RM, Silva PL, Carnier J, Oyama LM, Tock L, Tufik S, de Mello MT and Damaso AR (2011) The Role of PAI-1 and Adiponectin on the Inflammatory State and Energy Balance in Obese Adolescents with Metabolic Syndrome. *Inflammation* 35(3):944-951.

Damaso AR, de Piano A, Sanches PL, Corgosinho F, Tock L, Oyama LM, Tock L, do Nascimento CM, Tufik S and de Mello MT (2011) Hyperleptinemia in obese adolescents deregulates neuropeptides during weight loss. *Peptides* 32(7):1384-1391.

de Lima Sanches P, de Mello MT, Elias N, Fonseca FA, de Piano A, Carnier J, Oyama LM, Tock L, Tufik S and Damaso AR (2010) Improvement in HOMA-IR is an independent predictor of reduced carotid intima-media thickness in obese adolescents participating in an interdisciplinary weight-loss program. *Hypertens. Res.* 34(2):232-238.

de Piano A, Tock L, Carnier J, Foschini D, Sanches Pde L, Correa FA, Oyama LM, do Nascimento CM, Lederman HM, Ernandes R, de Mello MT, Tufik S and Damaso A (2011) The role of nutritional profile in the orexigenic neuropeptide secretion in nonalcoholic fatty liver disease obese adolescents. *Eur. J. Gastroenterol. Hepatol.* 22(5):557-563.

de Piano A, Tock L, Carnier J, Oyama LM, Oller do Nascimento CM, Martinz AC, Foschini D, Sanches PL, Ernandes RM, de Mello MT, Tufik S and Damaso AR (2010) Negative correlation between neuropeptide Y/agouti-related protein concentration and adiponectinemia in nonalcoholic fatty liver disease obese adolescents submitted to a long-term interdisciplinary therapy. *Metabolism* 59(5):613-619.

Dyzma M, Boudjeltia KZ, Faraut B and Kerkhofs M (2010) Neuropeptide Y and sleep. *Sleep Med. Rev.* 14(3):161-165.

Elias CF, Saper CB, Maratos-Flier E, Tritos NA, Lee C, Kelly J, Tatro JB, Hoffman GE, Ollmann MM, Barsh GS, Sakurai T, Yanagisawa M and Elmquist JK (1998) Chemically defined projections linking the mediobasal hypothalamus and the lateral hypothalamic area. *J. Comp. Neurol.* 402(4):442-459.

Fu LY, Acuna-Goycolea C and van den Pol AN (2004) Neuropeptide Y inhibits hypocretin/orexin neurons by multiple presynaptic and postsynaptic mechanisms: tonic depression of the hypothalamic arousal system. *J. Neurosci.* 24(40):8741-8751.

Fuhlendorff J, Gether U, Aakerlund L, Langeland-Johansen N, Thogersen H, Melberg SG, Olsen UB, Thastrup O and Schwartz TW (1990) [Leu31, Pro34]neuropeptide Y: a specific Y1 receptor agonist. *Proc. Natl. Acad. Sci. U S A* 87(1):182-186.

Gehlert DR, Schober DA, Beavers L, Gadski R, Hoffman JA, Smiley DL, Chance RE, Lundell I and Larhammar D (1996) Characterization of the peptide binding requirements for the cloned human pancreatic polypeptide-preferring receptor. *Mol. Pharmacol.* 50(1):112-118.

Gehlert DR, Schober DA, Gackenheimer SL, Beavers L, Gadski R, Lundell I and Larhammar D (1997) [125I]Leu31, Pro34-PYY is a high affinity radioligand for rat PP1/Y4 and Y1 receptors: evidence for heterogeneity in pancreatic polypeptide receptors. *Peptides* 18(3):397-401.

Ghamari-Langroudi M, Colmers WF and Cone RD (2005) PYY3-36 inhibits the action potential firing activity of POMC neurons of arcuate nucleus through postsynaptic Y2 receptors. *Cell Metab* 2(3):191-199.

Grouzmann E, Meyer C, Burki E and Brunner H (2001) Neuropeptide Y Y2 receptor signalling mechanisms in the human glioblastoma cell line LN319. *Peptides* 22(3):379-386.

Hernandez EJ, Whitcomb DC, Vigna SR and Taylor IL (1994) Saturable binding of circulating peptide YY in the dorsal vagal complex of rats. *Am J. Physiol.* 266(3 Pt 1):G511-516.

Holliday ND, Michel MC and Cox HM (2004) NPY Receptor Subtypes and Their Signal Transduction, in *Neuropeptide Y and Related Peptides Springer 2004; 1:46–67* (Michel MC ed) pp 46-67, Springer, New York.

Jacques D and Abdel-Samad D (2007) Neuropeptide Y (NPY) and NPY receptors in the cardiovascular system: implication in the regulation of intracellular calcium. *Can. J. Physiol. Pharmacol.* 85(1):43-53.

Jacques D, Dumont Y, Fournier A and Quirion R (1997) Characterization of neuropeptide Y receptor subtypes in the normal human brain, including the hypothalamus. *Neuroscience* 79(1):129-148.

Kamiji MM and Inui A (2007) Neuropeptide y receptor selective ligands in the treatment of obesity. *Endocr. Rev.* 28(6):664-684.

Kastin AJ and Akerstrom V (1999) Nonsaturable entry of neuropeptide Y into brain. *Am. J. Physiol.* 276(3 Pt 1):E479-482.

Larhammar D (1996) Evolution of neuropeptide Y, peptide YY and pancreatic polypeptide. *Regul. Pept.* 62(1):1-11.

Lerch M, Kamimori H, Folkers G, Aguilar MI, Beck-Sickinger AG and Zerbe O (2005) Strongly altered receptor binding properties in PP and NPY chimeras are accompanied by changes in structure and membrane binding. *Biochemistry* 44(25):9255-9264.

Masoudi Kazemabad A, Jamialahmadi K, Moohebati M, Mojarrad M, Dehghan Manshadi R, Akhlaghi S, Ferns GA and Ghayour-Mobarhan M (2012) Neuropeptide Y Leu7Pro Polymorphism Associated With the Metabolic Syndrome and Its Features in Patients With Coronary Artery Disease. *Angiology.*

McLean LR, Buck SH and Krstenansky JL (1990) Examination of the role of the amphipathic alpha-helix in the interaction of neuropeptide Y and active cyclic analogues with cell membrane receptors and dimyristoyl-phosphatidylcholine. *Biochemistry* 29(8):2016-2022.

Melnick I, Pronchuk N, Cowley MA, Grove KL and Colmers WF (2007) Developmental switch in neuropeptide Y and melanocortin effects in the paraventricular nucleus of the hypothalamus. *Neuron* 56(6):1103-1115.

Mercer RE, Chee MJ and Colmers WF (2011) The role of NPY in hypothalamic mediated food intake. *Front Neuroendocrinol* 32(4):398-415.

Michel MC, Beck-Sickinger A, Cox H, Doods HN, Herzog H, Larhammar D, Quirion R, Schwartz T and Westfall T (1998) XVI. International Union of Pharmacology recommendations for the nomenclature of neuropeptide Y, peptide YY, and pancreatic polypeptide receptors. *Pharmacol. Rev.* 50(1):143-150.

Munhoz R, Lazaretti-Castro M, Dâmaso AR, Tufik S, Tock L, Oyama LM, Nascimento CMO, Mello MT and Dâmaso A (2012) Influence of visceral and subcutaneous fat in bone mineral metabolism in obese adolescents. *Arquivos Brasileiros de Endocrinologia e Metabologia* 56:12-18.

Nonaka N, Shioda S, Niehoff ML and Banks WA (2003) Characterization of blood-brain barrier permeability to PYY3-36 in the mouse. *J. Pharmacol. Exp. Ther.* 306(3):948-953.

Oyama LM, do Nascimento CM, Carnier J, de Piano A, Tock L, Sanches Pde L, Gomes FA, Tufik S, de Mello MT and Damaso AR (2010) The role of anorexigenic and orexigenic neuropeptides and peripheral signals on quartiles of weight loss in obese adolescents. *Neuropeptides* 44(6):467-474.

Parker MS, Sah R, Balasubramaniam A, Sallee FR, Zerbe O and Parker SL (2011) Non-specific binding and general cross-reactivity of Y receptor agonists are correlated and should importantly depend on their acidic sectors. *Peptides* 32(2):258-265.

Parker SL, Parker MS, Sah R, Balasubramaniam A and Sallee FR (2008) Pertussis toxin induces parallel loss of neuropeptide Y Y(1) receptor dimers and G(i) alpha subunit function in CHO cells. *Eur. J. Pharmacol.* 579(1-3):13-25.

Pronchuk N, Beck-Sickinger AG and Colmers WF (2002) Multiple NPY receptors Inhibit GABA(A) synaptic responses of rat medial parvocellular effector neurons in the hypothalamic paraventricular nucleus. *Endocrinology* 143(2):535-543.

Robinson RB (1996) Autonomic receptor--effector coupling during post-natal development. *Cardiovasc. Res.* 31 Spec No:E68-76.

Shamseddeen H, Getty JZ, Hamdallah IN and Ali MR (2011) Epidemiology and economic impact of obesity and type 2 diabetes. *Surg. Clin. North Am.* 91(6):1163-1172, vii.

Tatemoto K, Carlquist M and Mutt V (1982) Neuropeptide Y--a novel brain peptide with structural similarities to peptide YY and pancreatic polypeptide. *Nature* 296(5858):659-660.

Tock L, Prado WL, Caranti DA, Cristofalo DM, Lederman H, Fisberg M, Siqueira KO, Stella SG, Antunes HK, Cintra IP, Tufik S, de Mello MT and Damaso AR (2006) Nonalcoholic fatty liver disease decrease in obese adolescents after multidisciplinary therapy. *Eur. J. Gastroenterol. Hepatol.* 18(12):1241-1245.

Tritos NA, Vicent D, Gillette J, Ludwig DS, Flier ES and Maratos-Flier E (1998) Functional interactions between melanin-concentrating hormone, neuropeptide Y, and anorectic neuropeptides in the rat hypothalamus. *Diabetes* 47(11):1687-1692.

Wahlestedt C, Regunathan S and Reis DJ (1992) Identification of cultured cells selectively expressing Y1-, Y2-, or Y3-type receptors for neuropeptide Y/peptide YY. *Life Sci.* 50(4):PL7-12.

Zhang Y, Proenca R, Maffei M, Barone M, Leopold L and Friedman JM (1994) Positional cloning of the mouse obese gene and its human homologue. *Nature* 372(6505):425-432.

Zou C, Kumaran S, Walser R and Zerbe O (2009) Properties of the N-terminal domains from Y receptors probed by NMR spectroscopy. *J. Pept. Sci.* 15(3):184-191.

In: Neuropeptide Y
Editors: Steven L. Parker

ISBN: 978-1-62618-421-3
© 2013 Nova Science Publishers, Inc.

Chapter III

Neuropeptide Y and its Receptors: Molecular Structure and Pathophysiological Role in Food Intake and Energy Homeostasis

Hidesuke Kaji
Division of Physiology and Metabolism,
University of Hyogo, Japan

Abstract

Neuropeptide Y (NPY) is an orexigenic peptide of 36-amino acids originally isolated from porcine brain. NPY and family peptides, peptide YY and pancreatic polypeptide, have pleiotropic effects upon food intake and energy homeostasis through NPY receptor family. Among known 8 types of NPY receptors, Y1, Y2 and Y5 receptors are the most studied in humans. A lot of experimental and clinical evidence from in vivo studies using selective NPY receptor agonists or antagonists, and from genetic manipulation, suggests that Y1 and Y5 receptors are involved in orexigenic signals and that Y2 receptor contributes anorexigenic signals. NPY neurons in hypothalamic arcuate nucleus not only project to ventromedial hypothalamus and inhibit satiety, but also project to lateral hypothalamus and increase food intake. Despite significant effects of

NPY on food intake as well as energy homeostasis by pharmacological manipulations, knockdown of NPY gene does not result in dramatic changes of phenotypic obesity. NPY effects on energy balance may be redundant due to contributions of central signals especially by melanocortin and orexin, or of a variety of peripheral signals, including particularly those by leptin, ghrelin and insulin. On the other hand, the NPY Y2 receptor might be involved in visceral obesity by direct effects on adipose tissues under stress with high-caloric diet, and in the development of plaque instability by direct effects on vascular smooth muscle cells. Associations between NPY/NPY receptor single nucleotide polymorphisms (SNPs) and obesity/type 2 diabetes mellitus / dyslipidemia / atherosclerosis have been reported in some cohort studies. We have reported that Y2R SNPs were associated with serum high density lipoprotein cholesterol levels. This review focuses on recent evidence about the pathophysiological mechanisms of NPY action, and on evaluation of NPY and NPY receptor promise as prophylactic and therapeutic targets in metabolic disorders and atherosclerosis-related diseases.

Keywords: Neuropeptide Y (NPY) , NPY Y1 receptor (Y1R), Y2 receptor (Y2R), Y5 receptor (Y5R), molecular structure, food intake, energy homeostasis, pathophysiololology, single nucleotide polymorphism (SNP)

Introduction

Cardiovascular diseases such as ischemic heart disease and stroke are major causes of death. Atherosclerosis is underlying pathogenic cause of ischemic cardiovascular diseases, and is significantly connected to smoking, stress, obesity, diabetes, hypertension and dyslipidemia. Obesity, particularly visceral obesity, is common cause of diabetes, hypertension and dyslipidemia. Obesity is, therefore, an important therapeutic as well as prophylactic target for cardiovascular diseases. Obesity is not only due to the lifestyle of overeating and low activity, but also to the genetic and epigenetic factors. Excessive eating leads to the disruption of energy homeostasis, altering the balance of varying orexigenic and anorexigenic signals in the central nervous system as well as in the peripheral tissues. Neuropeptide Y (NPY) is a representative orexigenic peptide expressed in the hypothalamic and autonomic neurons, acting through its receptor family. The first section of this chapter is a summary of the molecular structure of NPY and NPY receptors, which is important in understanding their functions. Functional NPY receptors especially in human are Y1R, Y2R and Y5R expressed in the hypothalamus and peripheral tissues. NPY and its receptors play a key role in food intake. The second portion of this review mainly summarizes the accumulated evidence regarding the pathophysiological role for NPY and NPY receptors in the control of food intake based upon a large number of experiments using pharmacological and genetic manipulations. Moreover, NPY plays a pleiotropic role not only in the central nervous system but also in the peripheral tissues. In particular, recent studies regarding the effects of NPY on adipose tissues as well as vascular system are together shown in this section. Studies about association of human NPY peptide and NPY receptor gene polymorphisms and metabolic phenotypes are addressed in the final section.

1. Molecular structure of NPY and its receptors

1.1. NPY

NPY was originally isolated from the porcine brain in 1982 (Tatemoto et al. 1982). The human NPY gene is located on chromosome 7P15.1 (Baker et al. 1995; Takeuchi etal. 1986), and consists of 4 exons and 3 introns. The initially generated precursor peptide pre-pro-NPY has 97 amino acids. The post-translational processing includes removal of N-terminal signal peptide of 27 amino acids, followed by cleavage by prohormone convertase to NPY (1-39) and C-terminal part of NPY (CPON). NPY (1-39) is further cleaved by carboxypeptidase-like enzyme (CPE) to NPY (1-37), in which the C-terminus is finally amidated by peptidylglycine α-amidating monooxygenase (PAM) to generate mature NPY (1-36)(Grouzmann et al. 2005; Larhammar et al. 1987; Higuchi et al. 1988; Williams et al. 1998). The non-helical C-terminal pentapeptide (TRQRY, NPY(32-36)) is highly conserved in NPY and PYY peptides and is critical for binding to Y1R, Y2R and Y5R, while Q^{34} combined with I^{31} (as in NPY) does not support binding to the Y4R (which accepts 31-36 VTRQRY of PYY at a lower affinity) (Fuhlendorff et al., 1990a; Silva et al., 2002; Wood et al., 1977; Glover et al., 1983; Glover et al., 1984).

1.2. NPY Receptors

The Y1R was first described in 1985 (Wahlestedt et al. 1986) and the human Y1R (gene name NPYR) gene was cloned in 1992(Herzog et al. 1992). Receptor subtypes Y1R, Y2R, Y4R and Y5R (Michel et al., 1998) are the most widely expressed among the known 8 subtypes of NPY/PP receptors (Y receptors)(Larhammar and Salaneck, 2004). Common structure of these receptors is heptahelical, rhodopsin-like (with transmembrane pivots N, D, R, W, P, P and N in helices 1-7).

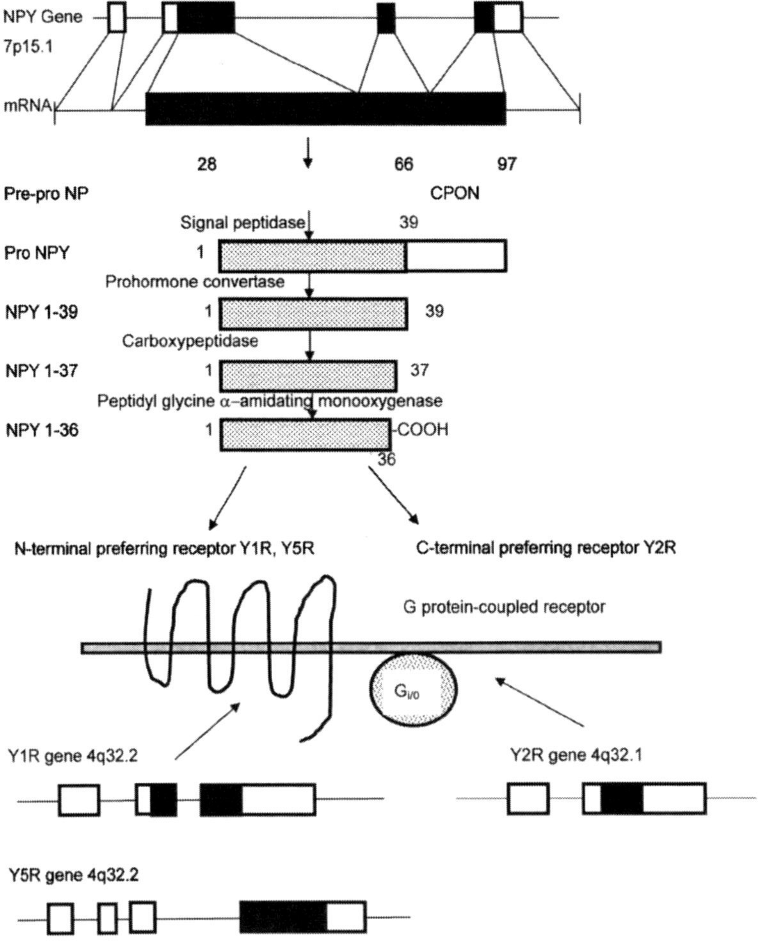

Figure 1. Molecular structure and biosynthesis of human NPY and its receptors.

These receptors are generated (Yasuda et al., 2009) and maintained (Parker et al., 2008a) as dimers coupled to G-protein heterotrimers (Baneres and Parello, 2003), with G_i-type (cyclase-inhibitory) α subunits as chief partners (Holliday et al., 2004). Ligand binding to Y1R, Y2R and Y5R (Moser et al., 2000) then primarily results in a lower cAMP formation through pertussis toxin-sensitive G protein subunits $G_{i/o}$. While the initial transduction after ligand binding to the receptor subtypes is common, each subtype has variable functions according to different affinities for PP family agonists, different expression in various cells, and different post-receptor downstream effectors such as intracellular calcium signaling proteins, phospholipases C and A2, mitogen-activated protein kinases, and nitric oxide-generating enzymes (Silva et al., 2002). Major subtypes of NPY receptors in human are the Y1R, Y2R, and Y5R. These NPY receptors are expressed in the hypothalamus, brain stem, and peripheral tissues such as blood vessels, lung, kidney, adrenal glands, stomach, colon, heart, pancreas and intestine.

1.2.1. The Y1 Receptor

The human Y1R gene has 8,831 base pairs (bp), consists of 2 coding exons and 1 non-coding exon and is located on chromosome 4q32.2. The length of the precursor mRNA is 2,974 bp and the mature receptor protein has 384 amino acids. NPY analogue, [Leu^{31}, Pro^{34}] NPY, is a ligand labeling all Y1 type receptors (Y1R, Y4R and Y5R), but has a much higher affinity for the Y1R relative to Y4R and Y5R subtypes. The C-terminal fragments NPY (3-36) and NPY(13-36) have very low affinity for the Y1R (Beck-Sickinger et al. 1994), indicating that this receptor requires a complete N-terminus of NPY and tolerates substitutions of the C-terminus of NPY(Inui A, 1999, Larhammar et al., 1992). Mutations Tyr100Ala in the 2nd extracellular domain or Phe286Glu at the end of the 6th transmembrane domain of the Y1R result in loss of ligand binding (Sautel et al., 1995; Sautel et al., 1996; Sjodin et al., 2006).

1.2.2. The Y2 Receptor

The Y2R gene has 8,448 bp, consists of 1 coding and 1 non-coding exon and is located on chromosome 4q32.1. The length of Y2R mRNA is 3,747 bp and the mature receptor protein has 381 amino acids. The Y2R structure is typical of rhodopsin-like heptahelical receptors. Y2 selective ligands peptide YY (3-36) (PYY (3-36)). NPY (3-36) and NPY(13-36) are full agonists, indicating that N-terminus of NPY is not required for specific binding to Y2R (Beck-Sickinger et al. 1994; Berglund et al., 2003). NPY(1-36), PYY(1-36),

NPY(3-36) and PYY(3-36) all have similar and very high affinity for the Y2R, while the affinity is much lower with shorter N-terminally clipped NPY PYY fragments, such as NPY(13-36), PYY(13-36), NPY(18-36) and NPY(22-36) (Parker and Balasubramaniam, 2008b). The Gln34Pro -modified NPY analog [Leu^{31}, Pro^{34}] NPY and PYY analog [Pro^{34}] PYY do not bind to the Y2R, indicating that Q34 is required for specific binding to Y2R (Fuhlendorff et al., 1990b; Potter and McCloskey, 1992; Kaga et al., 2001). At position 31, both isoleucine (in NPY) and valine (in PYY) allow high affinity binding to the Y2R. The Y1R however accepts both ITRPRY (in [Pro^{34}] NPY) and LTRPRY hexapeptide of mammalian pancreatic polypeptides (in [Leu^{31}Pro^{34}] NPY), but not the preceding 1-30 segment of PPs (Gehlert et al., 1997)

1.2.3. The Y5 Receptor

The human Y5R gene is located on human chromosome 4q31 in close association with the Y1R gene. The Y5R gene is transcribed from the promoter shared with the Y1R gene, in the opposite direction (Herzog et al., 1997). The gene consists of one coding and 3 non-coding exons, and is transcribed to precursor mRNA of 1,936 bp. The finished receptor protein consists of 445 amino acids, in which the third cytoplasmic domain is about 100 amino acids longer and C-tail is shorter than in other Y receptor subtypes (Berglund et al., 2003). The potency of ligand binding to Y5R is NPY > PYY = [Pro^{34}] NPY = NPY [2-36] ≥ PYY [3-36] > NPY[13-36], suggesting that the N-terminus of NPY is more adequate for specific binding to Y5R (Michel et al., 1998).

2. Pathophysiological Roles for NPY and its Receptors

Transducing through NPY receptor family, NPY plays multiple roles in food intake, locomotion, learning and memory, anxiety, epilepsy, pain, circadian rhythm, alcoholism, metabolic syndrome and cardiovascular functions.

This review focuses on the pathophysiological role for NPY and NPY receptors in the control of food intake and in the adipose tissues and vascular tone.

The central and peripheral signals, including those by NPY, are mainly integrated in specific hypothalamic areas to control food intake and energy homeostasis. Ventromedial hypothalamus (VMH) and dorsomedial hypothalamic nucleus (DMN) are satiety centers. Lateral hypothalamic area (LHA) and paraventricular nucleus (PVN) are feeding centers.

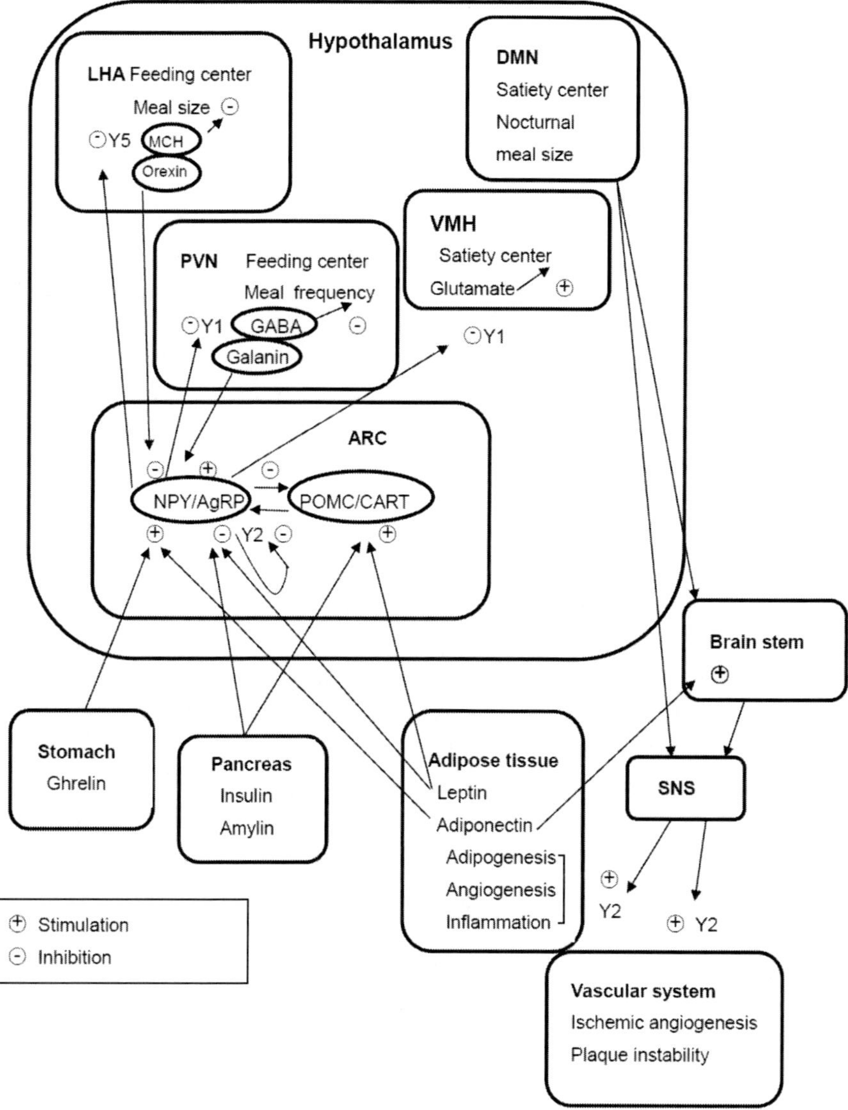

Figure 2. Role for NPY and its receptors in food intake and energy homeostasis.

NPY neurons in the arcuate nucleus (ARC) project to VMH, LHA and PVN, and those in DMN project to the brain stem and sympathetic nervous system (SNS).

2.1. Histology, Physiology and Feeding Pathology of NPY

NPY is expressed in whole brain as well as in peripheral nerves such as sympathetic neurons. The main source of NPY in the hypothalamus is ARC (Chronwall et al., 1985), projecting to PVN, VMH, and LHA. DMN is another source of NPY, projecting to the brain stem and SNS. The other abundant area of expression of NPY mRNA and immunoreactivity in the brain is hippocampus. In vivo application of NPY or similar agonist peptides into the third ventricle or discrete hypothalamic areas such as PVN, results in the increased food intake and body weight, and this could be reversed by pretreatment with Y1R or Y5R NPY antagonists, or by NPY antibodies (Wieland et al., 1998; Ishihara et al., 1998; Kanatani et al., 1996, 1998, 1999, 2001, Balasubramaniam et al., 2001; Rudolf et al., 1994; Morgan et al., 1998; Doods et al., 1995, 1996; Kask et al., 1998; Kannoa et al., 2001; Larsen et al., 1999).

NPY knockout mice of 129/Sv (Cp-J) strain is viable, fertile and does not display any major phenotypic difference when compared with NPY+/- mice (Erickson et al., 1996a).

NPY knockout mice of C57BL/6 strain surprisingly show mild obesity when compared with NPY+/+ control mice (Segal-Lieberman et al., 2003), suggesting the redundancy of other energy homeostatic mechanisms or the compensatory orexigenic drive by other factors at least during the development. When NPY knockout mice were crossed with leptin-deficient ob/ob mice, the phenotypic changes were reduced food intake, increased energy expenditure, and resistance to diet-induced obesity (DIO) (Erickson et al., 1996b).

Overexpression of NPY by an adeno-associated virus (AAV) in mouse hypothalamus resulted in obesity syndrome. The following findings illustrate NPY action in specific hypothalamic nuclei based on in vivo pharmacological evidence using selective NPY receptor agonists and antagonists (Table 1) as well as genetic manipulations of NPY.

2.1.1. NPY Actions in ARC

Male Wistar rats treated with an AAV expressing NPY anti-sense, or with RNA interference (RNAi) to cause 50% reduction of ARC NPY mRNA, showed reduced food intake and body weight (Gardiner et al., 2005). Transgenic mice overexpressing ARC NPY mRNA (by 18%) fed a palatable diet (with 50% energy from sucrose) develop obesity due to increased food intake (Takeuchi et al., 1986).

Table 1. Selective agonists and antagonists for NPY receptors Y1R, Y2R and Y5R

Receptor	Agonist	Antagonist
Y1R	$[Leu^{31}Pro^{34}]$ NPY	BIBO3304 #, LY357897 #
	$[Pro^{34}]$ NPY	J104870 #, 1229U91#
	$[Phe^7, Pro^{34}]$ NPY	BMS193885 (B), BIBP3226
	$[Leu^{31}, Pro^{34}]$ PYY	SR120107, SR120819A
	$[Pro^{34}]$ PYY	J-115814, H394184
Y2R	NPY(3-36), PYY(3-36)	BIIE0246 (B)
	BT-48, TASP-V	JNJ031020028 (B)
	PEGylated Y2 agonist	T4-[NPY 33-36]$_4$
	NPY(13-36),	
	Ac-$[Lys^{28}, Glu^{32}]$-NPY(25-36)	
Y5R	$[Ala^{31}, Aib^{32}]$ NPY	CGP71683A, MKL-0557
	$[cPP\ 1-7,\ NPY\ 19-23,\ Ala^{31},\ Aib^{32},\ Glu^{34}]$	PP Velneperit
	$[D-Trp^{34}]$ NPY	L-152,804
	$[K^4, RYYSA^{19-23}]$ PP(2-36)	α-(3-pyridylmethyl)-β-aminotetralin

(B): penetrant through the blood-brain-barrier

#: toxic to cardiovascular system when peripherally injected

2.1.2. NPY Actions in PVN

Among orexigenic effects, increased meal frequency by NPY could be caused by a stimulation of feeding center PVN (Pich et al., 1992; Stanley et al., 1985).

The mechanism is the suppression of inhibitory signals such as presynaptic GABA release from GABAergic terminals that synapse parvocellular neurons via Y1, 2 and 5 (Pronchuk et al., 2002) and the postsynaptic melanocortin 4 receptor (MC4R) neurons which could work through direct excitation of membrane hyperpolalization by Y1 (Ghamari-Langroudi et al., 2011; Gehlert et al., 1996). AAV-mediated overexpression of NPY in rat PVN caused an increase in meal frequency and a decrease in locomotion activity, leading to obesity and decreased insulin sensitivity (Tiesjema et al., 2007a, 2009).

2.1.3. NPY Actions in VMH

Intraneuronal NPY is less abundant in the satiety center VMH, and treatment with NPY has strong and direct inhibitory effects on this satiety center.

These effects could be mainly caused by a suppression of postsynaptic firing activity through inward rectifier K^+ channel (GirK) via Y1R (Chee et al., 2010). Suppression of presynaptic glutamatergic excitatory signals is more dominant than suppression of inhibitory GABAergic synaptic inputs to NPY-sensitive VMH neurons (Chance et al., 1991), and could also be involved.

2.1.4. NPY Actions in LHA

Sustained hyperphagia and weight gain with increased meal size could be caused by stimulation of the important feeding center, LHA. One of the mechanisms could be the suppression of inhibitory effects of melanin-concentrating hormone (MCH) through membrane hyperpolarization and inhibition of firing rate via Y5R (Van den Pol et al., 2004). Another mechanism could be the suppression of the negative feedback by LHA orexin on NPY mRNA expression and neuronal activity in ARC (De Lecea et al., 1998).

AAV-mediated overexpression of NPY in rat LHA has been reported to cause an increase in meal size and body weight (Tiesjema et al., 2007b).

2.1.5. NPY Actions in DMN

NPY expression levels in DMN of rodents are high in the early postnatal period during negative energy balance, and low thereafter (Grove et al., 2001, 2003a, b). Nocturnal meal size is increased by NPY injection in this satiety center.

The mechanism could involve DMN NPY neurons projecting to the brain stem and SNS.

In obese Otsuka-Long-Evans- Tokushima Fatty (OLETF) rats, intra-DMN treatment by NPY RNAi caused 40-45% decrease in DMN NPY mRNA and resulted in significant reduction of food intake, body weight and fat mass compared with controls (Yang et al., 2009). DMN NPY knockdown in rats was recently shown to promote the development of brown adipose tissue (BAT), thereby preventing DIO (Chao et al., 2011). AAV-mediated overexpression of NPY in rat DMN caused an increase in food intake (especially including high fat diet.) and body weight (Yang et al., 2009).

2.2. Histology, Physiology and Pathology of NPY Receptors as Related to Feeding

2.2.1. Feeding and the Y1 Receptor

Y1R mRNA expression in humans is high in ARC and PVN, layers II and VI of the neocortex, polymorphic layer of dentate gyrus, basal ganglia and amygdala (Caberlotto et al., 2000). Y1R-selective radioligands have low binding levels in the cortex, hypothalamus, and thalamus, and high binding levels in the stratum granulosum of the dentate gyrus of the hippocampus (Jacques et al., 1997). In the peripheral tissues, Y1R mRNA is abundant in rodent spleen, lung, liver, kidney, heart, skeletal muscle, and bone marrow (Nakamura et al., 1995) as well as in human colon, kidney, adrenal gland, and heart (Wharton et al., 1993).

Hypothalamic Y1R mRNA is increased in DIO-sensitive rats and reduced in DIO-resistant rats (Wang et al., 2007). Hypothalamic Y1R mRNA is increased by 25% in lean FA/FA rats compared with obese fa/fa rats (Beck et al., 2001). Induction of DIO in mice caused a reduction in VMN as well as DMN Y1R mRNA, leading to the inhibition of orexigenic signals (Zammaretti et al., 2007). Fasting and food deprivation caused a decrease in ARC (Cheng et al., 1998) and PVN (Zammaretti et al., 2001) Y1R mRNA, leading to the orexigenic drive.

Administration of the Y1R group agonists [Leu^{31}Pro^{34}] NPY or [Pro^{34}] NPY caused an increase in food intake in satiated rats (Stanley et al., 1992). Intracerebral injection of selective Y1R agonist caused an increase in both food intake and body weight, suggesting that Y1R is orexigenic receptor (Fekete et al., 2002; Lecklin et al., 2003; Mullins et al., 2001). On the other hand, Y1R antagonist BIBO3304 injected into the hypothalamus caused a reduction in fasting-induced hyperphagia (Wieland et al., 1998). Y1R antagonists LY357897, J-104870, and 1229U91 caused a reduction in NPY-induced food intake (Hipskind et al., 1997; Ishihara et al., 1998; Kanatani et al., 1999, 2001). Unfortunately, these antagonists injected systemically cannot pass the blood-brain-barrier and have adverse effects on the cardiovascular system. Peripheral injection of Y1R antagonist BMS-193885, which can enter the brain and is less toxic to heart, caused a reduction in both food intake and body weight (Antal-Zimanui et al., 2008). This compound may hold promise for treatment of severe obesity.

Food intake and fasting–induced hyperphagia is slightly reduced in Y1R knockout (Y1R-/-) mice. Reduction of food intake and fasting–induced hyperphagia are also observed in fed ad libitum Y1R, NPY -/- mice and NPY-treated Y1R, NPY -/- mice (Pedrazzini et al., 1998). When Y1R-/- mice is crossed with ob/ob mice, obesity is markedly attenuated via reduced food intake (Pralong et al., 2002). Unexpectedly, Y1R-/- mice of 129/Sv and C57BL/6 revealed late onset obesity without hyperphagia, suggesting more prominent compensatory hypometabolic responses related to age (Pedrazzini et al., 1998; Baldock et al., 2007; Kushi et al., 1998; Sainsbury et al., 2006).

2.2.2. Feeding and the Y2 Receptor

Human Y2R mRNA expression in the brain by in situ hybridization is highest in ARC, and also high in hippocampus, striatum, nucleus accumbens where Y1R mRNA colocalizes (Gehlert et al., 1996; Ammar et al., 1996; Gerald et al., 1995; Rose et al., 1995). Selective Y2R radioligands revealed high binding levels in medial preoptic area (MPOA) (Jacques et al., 1997). Y2R mRNA is also expressed in peripheral tissues such as kidney, heart, lung, liver, pancreas, spleen, bone marrow and thymus.

Y2R could be inhibitory presynaptic receptors on NPY neurons. ARC Y2R mRNA fails to change in fasting rats (Cheng et al., 1998). Hypothalamic Y2R mRNA is increased in DIO rats and decreased in DIO resistant (DIO-R) rats (Wang et al., 2007). While hypothalamic Y2R mRNA fails to change in DIO mice, it is reduced in the ARC of DIO-R mice (Huang et al., 2003).

Peripheral injection of Y2R agonist PYY (3-36) caused a reduction in food intake in humans and rodents, and of body weight in obesity models (Batterham et al., 2002; Degen et al., 2005; Pittner et al., 2004; Vrang et al., 2006).

Injection of the synthetic Y2R agonist BT-48 caused a reduction in food intake and an increase in fat metabolism in mice (Balasubramaniam et al., 2007). PEGylated Y2R agonist caused reductions in food intake and body weight, and improved glucose tolerance in mice (DeCarr et al., 2007; Lumb et al., 2007; Ortiz et al., 2007). These findings support an anorexigenic role for the Y2R.

Administration of Y2R antagonist BIIE0246 (which poorly traverses the BBB) increased feeding in satiated rats and decreased PYY-mediated reduction of feeding (Scott et al., 2005), suggesting that activation of he Y2R suppresses NPY neuronal activities and food intake. The BBB-penetrating Y2R antagonist JNJ031020028 normalizes food intake in stressed animals without affecting basal food intake (Shoblock et al., 2010). This compound inhibits stress-induced corticosterone levels in rodents.

Y2R knockout mice develop mild obesity with increased adiposity and decreased energy expenditure (Naveilhan et al., 1999). Y2R-/- mice of 129/Sv-C57BL/6 strain showed the increased feeding response after fasting but revealed decreased body weight and adiposity (Sainsbury et al., 2002). The discrepancy may be explained by different strains and targeting methods. Conditional knockdown of ARC Y2R revealed a transient increase in food intake with reduced body weight and no further increase in food intake after 4 weeks (Sainsbury et al., 2002). Knockdown of the Y2R gene in NPY-producing neurons caused an increase in food intake and body weight, supporting an anorexigenic role for the Y2R.

2.2.3. Feeding and the Y5 Receptor

Y5R mRNA in human brain is detected in the hypothalamus, ARC being particularly enriched. Strong hybridization signal is also noted in the stratum granulosum of the dentate gyrus and other regions of hippocampal formation, but not in the cerebral cortex, basal ganglia and thalamus (Hu et al., 1996; Weinberg et al., 1996; Statnick et al., 1998). Y5R mRNA is also found in peripheral tissues such as spleen, bone marrow, kidney, heart, skeletal muscle, lung, liver, pancreas, and thymus (Rodriguez et al., 2003).

Hypothalamic Y5R binding and mRNA are reduced in obese Zucker rats compared with lean controls (Beck et al., 2001). VMH, ARC and DMN Y5R mRNA are reduced in ob/ob mice (Xin and Huang, 1998). Hypothalamic Y5R mRNA is increased by DIO (Huang et al., 2003; Wang et al., 2007; Widdowson et al., 1997) and decreased by calorie restriction in mice and rats (Widdowson et al., 1997).

Y5R agonists [Ala^{31}, Aib^{32}] pNPY, [cPP$^{1-7}$, NPY19-23, Ala^{31}, Aib^{32}, Gln^{34}] hPP, [DTrp^{34}] hNPY, and [K^4, RYYSA$^{19-23}$] PP 2-36 caused an increase in food intake in the rat (Cabrele et al., 2000; McCrea et al., 2000; Parker et al., 2000).

Y5R antagonists CGP71683A caused a reduction in fasting or NPY-induced food intake (Criscione et al. 1998). However, this compound also blocks muscarinic receptors and serotonin reuptake (Della Zuana et al., 2001).

Clinical trial of Y5R antagonist MKL-0557 indicated a minimal reduction of body weight and did not enhance anti-obesity effects of orlistat or sibutramine (Erondu et al., 2006, 2007a,b). Clinical testing of Y5R antagonist velneperit induced modest weight loss, decreased waist circumstance and improved serum lipid profile (Shionogi and Co., 2009).

Y5R-/- mice showed mild obesity with increased food intake after 6 months (Marsh et al., 1998), but other Y5R-/- mice did not show any phenotypic change (Kanatani et al., 2000). Taken together, the Y5R does not appear to have a dominant role in NPY-mediated feeding.

3. Central and Peripheral Peptidergic Contributions Affecting NPY-Induced Food Intake

3.1. Central Peptidergic Signals

NPY-induced food intake is synergized by orexin, galanin (GAL) family peptides, and endocannabinoids. On the other hand, ARC NPY activity is inhibited by ARC melanocortin via MC4R and MC3R, and inhibited as autofeedback by ARC NPY via Y2R and by AgRP (agouti gene-related protein) via MC4R and MC3R (Fong et al., 1997).

3.1.1. Orexin

The orexin input from LHA stimulates food intake at least in part through the increased ARC NPY tone by increasing NPY mRNA and neuronal activity, which is supported by finding that Y1R and Y5R antagonists did not completely abolish orexin-induced food intake (Jain et al., 2000; Yamanaka et al., 2000; Dube et al., 2000). NPY from ARC causes a feedback inhibition of orexin neurons in LHA (Moreno et al., 2005).

3.1.2. Galanin Family

The galanin family includes galanin (GAL), GAL-like peptide (GALP) and a peptide produced by alternative splicing of the GALP transcript, alarin. These peptides are localized in various brain areas, including DMN, PVN and ARC (Melander et al., 1986; Skofitsh and Jacobowitz, 1985). GAL increases NPY from hypothalamic neurons cultured in vitro (Bergonzelli et al., 2001). A low dose of GAL (0.1 nM) does not affect food intake but blocks the Y1R agonist-induced food intake. A high dose of GAL (3nM) stimulates food intake but does not affect Y1R agonist-induced food intake (Parrado et al., 2007). These results suggest that GAL induces food intake by indirect activation of NPY system.

GALP is localized in ARC, median eminence and pituitary (Cunningham et al., 2002; Kerr et al., 2000). Acute GALP injection induced food intake in rats more prominently than GAL, which was however not present after 24h. GALP stimulates DMN NPY neurons, which is antagonized by Y1R antagonist BIBP3226 and 1229U91 (Kuramochi et al., 2006). Alarin stimulates food intake in male rats and causes c-fos expression in PVN, DMN, and ARC where NPY neurons exist (Van der Kolk et al., 2010). Addition of alarin stimulates NPY release from hypothalamic neurons cultured in vitro (Bopughton et al., 2010).

3.1.3. Endocannabinoids

Endocannabinoids are endogenous substance acting via cannabinoid receptor CB1 and CB2. Administration of endocannabinoids N-arachidonylethanolamine (AEA) (Jamshidi and Taylor, 2001) or 2-arachidonoylglycerol (Kirkham et al., 2002) into the hypothalamus stimulates food intake, which was blocked by CB1 antagonist rimonabant (Dodd et al., 2009).

Addition of AEA increased NPY release from the hypothalamic neurons cultured in vitro, which was blocked by the cannabinoid receptor blocker AM251 (Gamber et al., 2005). Administration of AEA caused a stimulation of NPY mRNA and protein in adult rats, which was blocked by rimonabant (Verty et al., 2009).

Thus, anorexic effect of CB1 blockade is due in part to the inhibition of NPY release in the hypothalamus.

3.1.4. Anorectic Melanocyte-Stimulating Hormones

Melanocyte-stimulating hormones (MSHs; also known as melanotropins and melanocortins) that also have anorectic activity include α-MSH (acetyl-ACTH(1-13)) and γ3-MSH (derived from the cryptic region of the ACTH/lipotropin precursor, pro-opiomelanocortin (POMC)). The ARC NPY provides a dominant inhibitory tone onto ARC POMC (Breen et al., 2005), and increases expression of agouti-related peptide (AgRP; an endogenous MSH antagonist) (Cone, 2005; Cowley, 2003a; Ollmann et al., 1997). POMC neurons are innervated by NPY-ergic terminals (Broberger et al., 1998) and express the Y1R (Hahn et al., 1998). A decrease in the release of γ3- MSH and an increased release of the AgRP were seen in fasting rats (Breen et al., 2005; Li et al., 2000).

3.2. Peripheral Signals

NPY-induced food intake is stimulated by ghrelin and adiponectin. On the other hand, leptin, insulin and amylin (Davidowa et al., 2004) inhibits ARC NPY activity, causing a reduction in their action potential frequency.

3.2.1. Ghrelin

Ghrelin is an acylated peptide generated in gastric X/A enteroendocrine cells (Date et al., 2000), providing hunger signal to the brain through parasympathetic nerves (Nakazato et al., 2001; Tschop et al., 2000; Wren et al., 2001a, b). Peripheral and central administration of ghrelin causes a stimulation of food intake and NPY/AGRP mRNA expression (Asakawa et al., 2001; Kamegai et al., 2001). Treatment by Y1R antagonist (Keen-Rhinenhart and Bartness, 2007) or NPY/AGRP knockdown (Chen et al., 2004) abolish ghrelin-induced food intake. Ghrelin stimulates ARC NPY neuronal activity by depolarization and increased firing frequency, thereby causing indirect inhibition of ARC POMC neuronal activities (Kohno et al., 2003, 2008). These effects are blocked by leptin administration (Kohno et al., 2007) or DIO (Briggs et al., 2010). Ghrelin also mimics NPY action in PVN by stimulating NPY release (Cowley et al., 2003b).

3.2.2. Adiponectin

Adiponectin is synthesized in white adipose tissues. Circulating adiponectin acts on ARC through subtype 1 of the adiponectin receptor (Adipo R1) to stimulate AMP kinase activity and food intake (Kubota et al., 2007). Adiponectin depolarizes, through Adipo R1 and R2, NPY neurons in ARC and solitary tract nucleus (NTS) of the brain stem, thereby stimulating NPY neuronal activity (Hoyda et al., 2009). Thus, NPY should be sensing adipose stores by adiponectin.

3.2.3. Leptin

Leptin is an anorexigenic peptide generated by white adipose cells. Leptin deficiency causes severe hyperphagic obesity with reduced satiety in rodents (Zhang et al., 1994) and humans (Montague et al., 1997), which is improved by leptin administration that reduces food intake and body weight (Campfield et al., 1995; Farooqi et al., 2001; Halaas et al., 1995; Pelleymounter et al., 1995).

Long form of leptin receptor LepRb is located in ARC, VMH, and DMN (Elmquist et al., 1998) where NPY neurons express LepRb (Baskin et al., 1999; Cowley et al., 2001). NPY mRNA levels are inversely related to circulating leptin levels (Korner et al., 2001). Leptin administration causes membrane hyperpolarization and reduces firing rate (Takahashi and Cone, 2005; Van den Top et al., 2004), thereby causing disinhibition of POMC neurons. Leptin also antagonizes ghrelin effects on NPY neurons that induce food intake (Kohno et al., 2008).

3.2.4. Insulin and Amylin

Both insulin and amylin are generated from pancreatic β cells. Administration of insulin into the brain causes acute reduction of food intake, especially of meal size (Air et al., 2002), and chronic reduction of body weight (Brief and Davis, 1984; Florant et al., 1991; Ikeda et al., 1986; Woods et al., 1979). Insulin receptors are expressed in ARC NPY neurons (Obici et al., 2002). Insulin reduces ARC NPY gene expression and NPY neurons in cultured ARC in vitro (Davidowa and Plagemann, 2007). Insulin is resistant in obesity and positive energy balance, and the inhibition of food intake and neuronal activity by insulin is not observed in such conditions (Davidowa and Plagemann, 2007). Peripheral and central administration of amylin causes reduction of food intake and chronic reduction of body weight (Chance et al., 1991; Morley and Flood, 1991; Lutz et al., 1994, 1998) through the synergistic inhibition with insulin of action potentials in ARC NPY neurons (Davidowa et al., 2004).

4. Actions of NPY in Adipose Tissue and Vascular System

4.1. Adipose Tissue

NPY directly activates proliferation and differentiation of preadipocytes, and indirectly activates fat growth by stimulation of angiogenesis in adipose tissues (Kuo et al., 2007). Y2R antagonist BIIE0246 inhibits this fat growth in vitro probably by increasing fat cell apoptosis as evidenced in fat pads of B6.V-Lep[ob]/J mice (Kuo et al., 2007). NPY also activates fat growth and angiogenesis in human fat xenografts transplanted to nude mice (Kuo et al., 2007).

High fat and sugar (HFS) diet plus chronic stresses (further abbreviated as HFS/ chronic stress) such as exposure to cold or aggressor increases plasma NPY levels and visceral fat as well as muscle and liver fat. NPY, Y2R and dipeptidyl peptidase IV (DPP4), which generates Y2-preferring form of NPY, NPY (3-36), are increased under such conditions. Visceral, muscle and liver fat induced by HFS/chronic stress is abolished in mice by Y2R antagonist or Y2R knockout (systemic or conditional). HFS/chronic stress causes an increase in plasma glucocorticoid levels and fat corticosteron levels by stimulation of hydroxysteroid dehydrogenase (HSD)11β1, but it is not abolished by Y2R antagonist. While HFS/chronic stress acutely stimulates sympathetic nerves and the release of catecholamines leading to lipolysis, chronical application of this treatment reduces catecholamine release and β-adrenergic lipolytic action. The HFS /chronic stress also stimulates NPY release from sympathetic nerve terminals and activates macrophage infiltration in fat tissues, thereby causing insulin resistance (Kuo et al., 2007).

These results suggest that HFS/chronic stress induce adipogenesis, angiogenesis and inflammation by NPY from sympathetic nerves in fat tissues by mediating active glucocorticoid induction as upstream signals.

DMN NPY knockdown in rats was shown previously to promote the development of BAT, which prevented DIO (Chao et al., 2011). NPY plays a role in the regulation of gene transcription through ERK and STAT3 pathway in vascular stromal cells from BAT but did not play any role in thermogenesis (Shimada et al., 2012).

4.2. Vascular System

NPY is present in sympathetic nerve endings around the blood vessels. Circulating NPY levels are increased by high sympathetic neuronal activities caused by cardiovascular diseases such as acute myocardial infarction, angina pectoris, heart failure and hypertension (Malmstrom and Lundberg, 2002).

4.2.1. Cardiovascular Effects

High frequency of sympathetic stimulation in guinea pig vena cava induces long-lasting vasoconstriction in vitro, which is reversed by Y1R antagonists SR120107 and BIBP3226 (Malmstrom and Lundberg, 1995a, b; Lundberg and Modin, 1995). Stress causes hypertension in pigs, which is recovered by Y1R antagonist BIBP3226 (Han et al., 1998; Wahlestedt et al., 1985).

These findings are substantiated by the evidence that chronic stress stimulates NPY release from adrenal medulla (Takiyyuddin et al., 1994). NPY potentiates noradrenaline-induced vasoconstriction through Y1R, suggesting a cross-talk between catecholamines and NPY (Qureshi et al., 1998). NPY maintains blood pressure during septic and hemorrhagic shock, which is cancelled by Y1R antagonist BIBP3226 (Wahlestedt and Hakanson, 1986a). On the other hand, Y2R inhibits noradrenaline release from SNS (Malmstrom, 2001). This effect is blocked by Y2R antagonist BIIE0246 (Kawamura et al., 1991).

4.2.2. Vascular Endothelial Cells

NPY stimulates prostacyclin secretion from porcine aortic endothelial cells (EC) (Fabi et al., 1998), indicating that NPY plays a role in endothelium-dependent regulation of vascular motility. Potentiation by NPY of noradrenaline-induced vasoconstriction is mediated by thromboxane released from endothelium of human saphenous vein (Noll et al., 1996). In rat microvascular coronary EC, NPY reduces permeability to macromolecules and intracellular cAMP content through the involvement of G protein Gi (Zukowska-Grojec et al., 1998). NPY causes EC proliferation, migration, and adherence to the extracellular matrix (Ghersi et al., 2001). It also stimulates capillary tube formation in vitro and angiogenesis in vivo (Ghersi et al., 2001). NPY induced EC migration in response to wounding, suggesting that selective Y2 agonists could be beneficial to wound healing (Pankajakshan et al., 2011).

4.2.3. Vascular Smooth Muscle Cells

Divya Pankajakshan et al. have reported that Y1R and Y5R mRNA are more abundant in isolated vascular smooth muscle cells (VSMC) of healthy carotid artery than symptomatic or asymptomatic plaque VSMC (Pankajakshan et al., 2011). In contrast, Y2R mRNA is more abundant in symptomatic or asymptomatic plaque VSMC than in healthy VSMC. Inflammatory cytokines such as TNFα, interleukin (IL)-12 and interferon-γ (IFNγ) caused a decrease in Y1R and Y5R mRNA and an increase in Y2R mRNA in healthy VSMC of carotid artery. These cytokines caused similar effects on Y2R and Y5R mRNA in plaque VSMC. An increase by cytokines in Y2R mRNA is more prominent in asymptomatic than symptomatic plaque VSMC, suggesting that symptomatic plaque is already exposed to endogenous cytokines.

Y1R mRNA is reduced by IFNγ in asymptomatic plaque VSMC and by IL-12 in symptomatic plaque VSMC. Previous studies indicated that Y1R and

Y5R are involved in mitogenesis of rat aortic VSMC. These results are supported by the evidence that antagonists of both Y1R and Y5R reduced the neointima formation induced by angioplasty (Lee et al., 2003a).

Y2R expression is low in healthy VSMCs and artery sections, but is high following NPY injection, injury and ischemia (Lee et al., 2003b). Y2R knockout diminished NPY-induced angiogenesis both in the aortic sprouting and in the ischemic limb model in mice (Wahlestedt et al., 1986b), suggesting a major role for the receptor in NPY-mediated angiogenesis.

Taken together, atheroma associated cytokines reduces Y1R and Y5R expression and increases Y2R expression in VSMCs of symptomatic carotid plaques, thereby leading to the reduced proliferation of plaque VSMC, with increased plaque angiogenesis and thus plaque instability as a risk for acute coronary syndrome.

5. Association between NPY and NPY receptor gene polymorphisms and metabolic traits

Table 2. Association between NPY/NPY receptor SNPs and metabolic traits

SNPs	Metabolic traits	Subjects
NPY Leu7Pro	birth weight, serum TG	preschool aged children
	carotid atherosclerosis	elderly patients with T2DM
	serum TC AND LDL-C	
	serum TC	women with CHD
PYY rs162430	obesity	children and women
-1746, -1653	obesity	Pima Indian
rs162430		
Y1R, Y5R C/T in untranslated region between Y1R and Y5R	serum TG and HDL-C	obese subjects
Y2R rs1047214, rs2880415	severe obesity	Pima Indian men
rs10472214	waist/hip ratio	severely obese children
rs2880415	obesity	men

585T>C	protective against obesity	Swedish Caucasian obese men
rs12649641	obesity	Danish white subjects
rs12649641 rs2880415 rs17376826	BMI (additive to FTO MC4R, NPFFR2)	Utah subjects

rs2880416
haplotypes
  CTCAGAA
  GTCAGAA
	BMI and T2DM	Scandinavian, Swedish, French Canadian, American European & Polish men

rs6857715	severe obesity	French white adult subjects French white children (modest)
rs6857530 rs6857715	serum HDL-C	Japanese subjects

T2DM: type 2 diabetes mellitus CHD: coronary heart disease TC: total cholesterol LDL-C: low density lipoprotein-cholesterol HDL-C: high density lipoprotein-cholesterol TG: triglyceride.

5.1. Obesity

Prepro NPY variant of signal peptide, Leu7Pro, is associated with birth weight (Karvonen et al., 2000). Birth weight is 193 g larger in subjects with Pro7 compared to those with Leu7. Variants of PYY rs162430 are associated with child or female obesity (Hung et al., 2004). NPY and Y1R variant are associated with morbid obesity in linkage study of French Caucasian families (Roche et al., 1997). In contrast, cohorts from different regional backgrounds did not support association of Y1R and Y5R genes with obesity. In a study of obese Pima Indian, 3 haplotypes of the Y2R gene are modestly associated with severe obesity (Ma et al., 2005). Y2R SNPs rs10472214 (585C/T) and rs2880415 (936C/T) are associated with obesity only in men (Hung et al., 2004). From a study of Swedish Caucasian obese men, a common Y2R variant should have a role against obesity (Lavebratt et al., 2006). Y2R SNP rs6857715 is associated with severe adult obesity and modestly with childhood obesity, and rs1047214 is associated with waist / hip ratio in the severely obese children (Siddiq et al., 2007). Y2R SNP rs12649641 is associated with obesity in Danish white subjects (Torekov et al., 2006). More recently, polymorphisms in the Y2R gene SNPs rs12649641, rs2880415, and rs17376826 were found to have significant associations with BMI that are additive to FTO, MC4R, and NPFFR2 gene effects in Utah subjects (Hunt et al., 2011). Haplotypes of Y2R, NPFFR2 and MC4R, plus FTO SNP, explained 9.6% of the BMI variance in this study.

5.2. Diabetes Mellitus

Pre-pro-NPY variant of signal peptide, Leu7Pro, is associated with carotid atherosclerosis in elderly patients with type 2 diabetes mellitus (T2DM) (Niskanen et al., 2000). An Y2R variant is associated with BMI and T2DM (Cambell et al., 2007).

5.3. Dyslipidemia

Polymorphisms of leucine 7 to proline 7 in the pre-pro NPY are associated with high serum cholesterol and low density lipoprotein cholesterol levels (Karvonen et al., 1998), and with higher serum cholesterol levels in women with coronary heart disease (Erkkila et al., 2002). Polymorphisms of leucine 7 to proline 7 in the pre-pro NPY are associated with serum triglyceride (TG) levels in boys of preschool age (Wahlestedt et al., 1985). Y1R and Y5R variants are associated with serum TG and high density lipoprotein-cholesterol (HDL-C) levels (Blumenthal et al., 2002). We have previously reported that Y2R SNPs rs6857715 and rs6857530 are associated with serum HDL-C levels in men (Takiguchi et al., 2010). Serum HDL-C levels are significantly and independently lower in subjects with TT than TC or CC of Y2R SNP rs6857715 and lower in subjects with GG than those with GA or AA of Y2R SNP rs6857530. Both SNPs are located on the promoter region of the Y2R gene and thus may affect Y2R transcriptional activity.

Summary and Conclusion

Accumulating evidence suggests that hypothalamic NPY plays a key orexigenic role in the control of food intake through Y1R by integrating central and peripheral energy signals. In addition to the central action of NPY, there are interesting findings regarding peripheral action of NPY. In animals under stress that are fed high glucose and fat diet, NPY from peripheral sympathetic neurons induces adipogenesis, angiogenesis, and inflammation through Y2R in adipose tissues. In human studies, Y2R expression in VSMC of symptomatic carotid plaques may point to plaque instability as a cause of acute coronary syndrome. In concordance with these findings, some NPY/NPY receptor SNPs are associated with obesity, diabetes and dyslipidemia. Many NPY receptor specific agonists or antagonists have been synthesized based upon the molecular structures shown in Table 1. NPY has a variety of effects on systemic tissues, and most NPY agonists or antagonists are beneficial for some pathologic state but not for the other. Thus, an appropriate matching of receptor subtypes and site-selective NPY agents is necessary to prevent and treat metabolic and cardiovascular diseases that relate to NPY, and this of course is also critical in developing and testing new NPY system-active drugs.

References

Air EL, Benoit SC, Blake Smith KA, Clegg DJ, and Woods SC (2002) Acute third ventricular administration of insulin decreases food intake in two paradigms. *Pharmacol. Biochem. Behav.* 1-2: 423-429.

Ammar DA, Eadie DM, Wong DJ, Ma YY, Kolakowski Jr LF, Tang-Feng TL, and Thompson DA (1996) Characterization of human type 2 neuropeptide Y receptor gene (NPY2R) and localization to the chromosome 4q region containing the type 1 neuropeptide Y receptor gene. *Genomics* 3: 392-398.

Antal-Zimanyi I, Bruce MA, Leboulluec KL, Iben LG, Mattson GK, McGovern RT, Hogan JB, Leahy CL, Flowers SC, Stanley JA, Ortiz AA, and Poindexter GS (2008). Pharmacological characterization and appetite suppressive properties of BMS-193885, a novel and selective neuropeptide Y (1) receptor antagonist. *Eur. J. Pharmacol.* 1-3: 224-232.

Asakawa A, Inui A, Kaga T, Yuzuriha H, Nagata T, Ueno N, Makino S, Fujimiya M, Niijima A, Fujino MA, and Kasuga M (2001) Ghrelin is an appetite-stimulatory signal from stomach with structural resemblance to motilin. *Gastroenterology* 2: 337-345.

Baker E, Hort YJ, Ball H, Sutherland GR, Shine J, and Herzog H (1995) Assignment of the human neuropeptide Y gene to chromosome 7p15.1 by nonisotopic in situ hybridization. *Genomics* 26:163-164.

Balasubramaniam A, Dhawan VC, Mullins DE, Chance WT, Sheriff S, Guzzi M, Prabhakaran M, and Parker EM (2001) Highly selective and potent neuropeptide Y (NPY) Y1 receptor antagonists based on [Pro(30),

Tyr(32), Leu(34)]NPY(28–36)-NH2 (BW1911U90). *J. Med. Chem.* 44: 1479–1482.

Balasubramaniam A, Joshi R, Su C, Friend LA, and James JH (2007) Neuropeptide Y (NPY) Y2 receptor-selective agonist inhibits food intake and promotes fat metabolism in mice: combined anorectic effects of Y2 and Y4 receptor-selective agonists. *Peptides* 2: 235-240.

Baldock PA, Allison SJ, Lundberg P, Lee NJ, Slack K, Lin EJ, Enriquez RF, McDonald MM, Zhang L, During MJ, Little DG, Eisman JA, Gardiner EM, Yulyaningsih E, Lin S, Sainsbury A, and Herzog H (2007) Novel role of Y1 receptors in the coordinated regulation of bone and energy homeostasis. *J. Biol. Chem.* 26: 19092-19102.

Banères JL, Parello J (2003) Structure-based analysis of GPCR function: evidence for a novel pentameric assembly between the dimeric leukotriene B4 receptor BLT1 and the G-protein. *J. Mol. Biol.* 329:815-829.

Baskin DG, Breininger JF, Bonigut S, and Miller MA (1999) Leptin binding in the arcuate nucleus is increased during fasting. *Brain Res.* (1-2)154-158.

Batterham RL, Cowley MA, Small CJ, Herzog H, Cohen MA, Dakin CL, Wren AM, Brynes AE, Low MJ, Ghatei MA, Cone RD, and Bloom SR (2002) Gut hormone PYY(3-36) physiologically inhibits food intake. *Nature* 6898: 650-654.

Berglund MM, Hipskind PA, and Gehlert DR (2003) Recent developments in our understanding of the physiological role or PP-fold peptide receptor subtypes. *Exp. Biol. Med.* (Maywood) 228: 217-244.

Bergonzelli GE, Pralong FP, Glauser M, Cavadas C, Grouzmann E, and Gaillard RC (2001) Interplay between galanin and leptin in the hypothalamic control of feeding via corticotrophin-releasing hormone and neuropeptide Y. *Diabetes* 12: 2666-2672.

Beck B, Richy S, Dimitrov T, and Stricker-Krongrad A (2001) Opposite regulation of hypothalamic orexin and neuropeptide Y receptors and peptide expressions in obese Zucker rats. *Biochem. Biophys. Res. Commun.* 3: 518-523.

Beck-Sickinger AG, Wieland HA, Wittneben H, Willim KD, Rudolf K, and Jung G (1994) Complete L-alanine scan of neuropeptide Y reveals ligands binding to Y1 and Y2 receptors with distinguished conformations. *Eur. J. Biochem.* 225:947-958.

Blumenthal JB, Andersen RE, Mitchel BD, Seibert MJ, Yang H, Herzog H, Beamer BA, Franckowiak SC, and Walston JD (2002) Novel neuropeptide Y1 and Y5 receptor gene variants: associations with serum triglyceride and high-density lipoprotein cholesterol levels. *Clin. Genet.* 62: 196-202.

Boughton CK, Patterson M, Bewick GA, Tadross JA, Gardiner JV, Beale KE, Chaudery F, Hunter G, Busbridge M, Leavy EM, Ghatei MA, Bloom SR, and Murphy KG (2010) Alarin stimulates food intake and gonadotrophin release in male rats. *Br. J. Pharmacol.* 3: 601-613.

Breen TL, Conwell IM, and Wardlaw SL (2005) Effects of fasting, leptin, and insulin on AGRP and POMC peptide release in the hypothalamus. *Brain Res.* 1032: 141-148.

Brief DJ, and Davis JD (1984) Reduction of food intake and body weight by chronic intraventricular insulin infusion. *Brain Res. Bull.* 5: 571-575.

Briggs DI, Enriori PJ, Lermus MB, Cowley MA, and Andrews ZB (2010) Diet-induced obesity causes ghrelin resistance in arcuate NPY/AgRP neurons. *Endocrinology* 10: 4745-4755.

Broberger C, De Lecea L, Sutcliffe JG, and Hockfelt T (1998) Hypocretin/orexin- and melanin concentrating hormone-expressing cells form distinct populations in the rodent lateral hypothalamus: relationship to the neuropeptide Y and agouti gene-related protein systems. *J. Comp. Neurol. 402*: 460-474.

Caberlotto L, Fuxe K, and Hurd YL (2000) Characterization of NPY mRNA-expressing cells in the human brain: co-localization with Y2 but not Y1 mRNA in the cerebral cortex, hippocampus, amygdala, and striatum. *J. Chem. Neuroanat.* 20: 327-337.

Cabrele C, Langer M, Bader R, Wieland HA, Doods HN, Zerbe O, and Beck-Sickinger AG (2000) The first selective agonist for the neuropeptide Y Y5 receptor increases food intake in rats. *J. Biol. Chem.* 46: 36043-36048.

Cambell CD, Lyon HN, Nemesh J, Drake JA, Tuomi T, Gaudet D, Zhu X, Cooper RS, Ardlie KG, Groop LC, Hirschhorn JN(2007) Association studies of BMI and type 2 diabetes in the neuropeptide Y pathway: a pssible role for NPY2R as a candidate gene for type 2 diabetes in men. *Diabestes* 56: 1460-1467

Campfield LA, Smith FJ, Guisez Y, Devos R, and Burn P (1995) Recombinant mouse OB protein: evidence for a peripheral signal linking adiposity and central neural networks. *Science* 5223: 546-549.

Chance WT, Balasubramanjam A, Zhang FS, Wimalawansa SJ, and Fischer JE (1991) Anorexia following the intrahypothalamic administration of amylin. *Brain Res.* 2: 352-354.

Chao PT, Yang L, Aja S, Moran TH, and Bi S (2011) Knockdown of NPY expression in the dorsomedial hypothalamus promotes development of brown adipocytes and prevents diet-induced obesity. *Cell Metab.* 5: 573-583.

Chee MJ, Myers Jr MG, Price CJ, and Colmers WF (2010) Neuropeptide Y suppresses anorexigenic output from the ventromedial nucleus of the hypothalamus. *J. Neurosci.* 9: 3380-3390.

Chen HY, Trumbauer ME, Chen AS, Weingarth AT, Adams JR, Fraizier EG, Shen Z, Marsh DJ, Feighner SD, Guan XM, Ye Z, Nargund RP, Smith RG, Van der Ploeg LH, Howard AD, MacNeil DJ, and Quian S (2004) Orexigenic acton of peripheral ghrelin is mediated by neuropeptide Y and agouti-related protein. *Endocrinology* 6: 2607-2612.

Cheng X, Broberger C, Tong Y, Yongtao X, Ju G, Zhang X, and Hokfelt T (1998) Regulation of expression of neuropeptide Y Y1 and Y2 receptors in the arcuate nucleus of fasted rats. *Brain Res.* 1: 89-96.

Chronwall BM, DiMaggio DA, Massari VJ, Pickel VM, Ruggiero DA, and O'Donohue TL (1985) The anatomy of neuropeptide-Y-containing neurons in rat brain. *Neurosci.* 4: 1159-1181.

Cone RD (2005) Anatomy and regulation of the central melanocortin system. *Nat. Neurosci.* 8: 571-578.

Cowley MA, Smart JL, Rubinstein M, Cerdan MG, Diano S, Horvath TL, Cone RD, and Low MJ (2001) Leptin activates anorexigenic POMC neurons through a neural network in the arcuate nucleus. *Nature* 6836: 480-484.

Cowley MA (2003a) Hypothalamic melanocortin neurons integrate signals of energy state. *Eur. J. Pharmacol.* 480: 3-11.

Cowley MA, Smith RG, Diano S, Tshop M, Pronchuk N, Grove KL, Strasburger CJ, Bidingmaier M, Esterman M, Heiman ML, Garcia-Segura LM, Nillni EA, Mendez P, Low MJ, Sotonyi P, Friedman JM, Liu H, Pinto S, Colmers WF, Cone RD, and Horvath TL (2003b) The distribution and mechanism of action of ghrelin in the CNS demonstrates a novel hypothalamic circuit regulating energy homeostasis. *Neuron* 4: 649-661.

Criscione L, Rigollier P, Batzl-Hartmann C, Rueger H, Stricker-Krongrad A, Wyss P, Brunner L, Whitebread S, Yamaguchi Y, Gerald C, Heurich RO, Walker MW, Chiesi M, Schilling W, Hofbauer KG, and Levens N (1998) Food intake in free-feeding and energy-deprived lean rats is mediated by the neuropeptide Y5 receptor. *J. Clin. Invest.* 12: 2136-2145.

Cunningham MJ, Scarlett JM, and Steiner RA (2002) Cloning and distribution of galanin-like peptide mRNA in the hypothalamus and pituitary of the macaque, *Endocrinology* 3: 755–763.

Davidowa H, Ziska T, and Plagemann A (2004) Arcuate neurons of overweight rats differ in their responses to amylin from controls. *Neuroreport* 18: 2801-2805.

Davidowa H, Li Y, and Plagemann A (2007) Insulin resistance of hypothalamic arcuate neurons in neonatally overfed rats. *Neuroreport* 5: 521-524.

Date Y, Kojima M, Hosoda H, Sawaguchi A, Mondal MS, Suganuma T, Matsukura S, Kangawa K, and Nakazato M (2000) Ghrelin, a novel growth hormone-releaseing acylated peptide, is synthesized in a distinct endocrine cell type in the gastrointestinal tracts of rats and humans. *Endocrinology* 11: 4255-4261.

DeCarr LB, Buckholz TM, Milardo LF, Mays MR, Ortiz A, and Lumb KJ (2007) A long acting selective neuropeptide Y2 receptor PEGylated peptide agonist reduces food intake in mice. *Bioorg. Med. Chem. Lett.* 7: 1916-1919.

De Lecea L, Kilduff TS, Peyron C, Gao X, Foye PF, Danielson PE, Fukuhara C, Battenberg EL, Gautvik VT, Barlett 2nd FS, Frankel WN, van den Pol AN, Bloom FE, Gatvik KM, and Sutcliffe JG (1998) The hypocretins: hypothalamic-specific peptides with neuroexcitatory activity. *Proc. Natl. Acad. Sci. USA* 1: 322-327.

Degen L, Oesch S, Casanova M, Graf S, Ketterer S, Drewe J, and Beglinger C (2005) Effect of peptide YY3-36 on food intake in humans. *Gastroenterology* 5: 1430-1436.

Della Zuana O, Sadio M, Germain M, Feletou M, Chamorro S, Tisserand F, de Montrion C, Boivin JF, Duhault J, Boutin JA, and Levens N (2001) Reduced food intake in response to CGP71683A may be due to mechanisms other than NPY Y5 receptor blockade. *Int. J. Obes. Relat. Metab. Disord.* 1: 84-94.

Dodd GT, Stark JA, McKie S, Williams SR, and Luckman SM (2009) Central cannabinoid signaling mediating food intake: a pharmacological-challenge magnetic resonance imaging and functional histology study in rat. *Neuroscience* 4: 1192-1200

Doods HN, Wienen W, Entzeroth M, Rudolf K, Eberlein W, Engel W, and Wieland HA (1995) Pharmacological characterization of the selective nonpeptide neuropeptide Y Y1 receptor antagonist BIBP 3226. *J. Pharmacol. Exp. Ther.* 275: 136–142.

Doods HN, Wieland HA, Engel W, Eberlein W, Willim KD, Entzeroth M, Wienen W, and Rudolf K (1996) BIBP 3226, the first selective neuropeptide Y1 receptor antagonist: a review of its pharmacological properties. *Regul. Pept.* 65: 71–77.

Dube MG, Horvath TL, Kalra PS, and Karla SP (2000) Evidence of NPY Y5 receptor involvement in food intake elicited by orexin A in sated rats. *Peptides* 10: 1557-1560.

Erickson JC, Clegg KE, and Palmiter RD (1996a) Sensitivity to leptin and susceptibility to seizures of mice lacking neuropeptide Y. *Nature* 6581: 415-421.

Erickson JC, Hollopeter G, and Palmiter RD (1996b) Attenuation of the obesity syndrome of ob/ob mice by the loss of neuropeptide Y. *Science* 5293: 1704-1707.

Elmquist JK, Bjorbaek C, Ahima RS, Flier JS, and Saper CB (1998) Distributions of leptin receptor mRNA isoforms in the rat brain. *J. Comp. Neurol.* 4: 535-547.

Erkkila AT, Lindi V, Lehto S, Laakso M, and Uusitpa MI (2002) Association of leucine 7 to proline 7 polymorphism in the preproneuropeptide Y with serum lipids in patients with coronary heart disease. *Mol. Genet. Metab.* 75: 260-264.

Erondu N, Gantz I, Musser B, Suryawanshi S, Malick M, Addy C, Cote J, Bray G, Fujioka K, Bays H, Hollander P, Sanabria-Bohorquez SM, Eng W, LangstromB, Hargreaves RJ, Burns HD, Kanatani A, Fukami T, MacNeil Dj, Gottesdiener KM, Amatruda JM, Kaufuman KD, and Heymsfield SB (2006) Neuropeptide Y5 receptor antagonism does not induce clinically meaningful weight loss in overweight and obese adults. *Cell Metab.* 4: 275-282.

Erondu N, Addy C, Lu K, MAllick M, Musser B, Gantz I, Proietto J, Asrup A, Toubro S, Rissannen AM, Tonstad S, Hynes WG, Gottesdiener KM, Kaufman KD, Amatruda JM, and Heymsfield SB (2007a) NPY5R antagonism does not augment the weight loss efficacy of orlistat or sibutramine in Obesity, pp2027-2042, *Silver Spring*.

Erondu N, Wadden T, Gantz I, Musser B, Nguyen AM, Bays H, Bray G, O'Neil PM, Basdevant A, Kaufaman KD, Heymsfield SB, and Amatruda JM (2007b) Effect of NPY5R antagonist MK-0557 on weight regain after very low-calorie diet-induced weight loss, in Obesity pp895-905, *Silver Spring*.

Fabi F, Argiolas L, Ruvolo G, and del Basso P (1998) Neuropeptide Y-induced potentiation of noradrenergic vasoconstriction in the human saphenous vein: involvement of endothelium generated thromboxane. *Br. J. Pharmacol.* 124: 101-110.

Farooqi IS, Keogh JM, Kamath S, Jones S, Gibson WT, Trussell R, Jebb SA, Lip GY, and O'Rahilly S (2001) Partial leptin deficiency and human adiposity. *Nature* 6859: 34-35.

Fekete C, Sarkar S, Rand WM, Harney JW, Emerson CH, Biance AC, Beck-Sickinger A, and Lechan RM (2002) Neuropeptide Y1 and Y5 receptors mediate the effects of neuropeptide Y on the hypothalamic-pituitary-thyroid axis. *Endocrinology* 12: 4513-4519.

Florant GL, Singer L, Scheurink AJ, Park CR, Richardson RD, and Woods SC (1991) Intraventricular insulin reduces food intake and body weight of marmots during the summer feeding period. *Physiol. Behav.* 2: 335-338.

Fong TM, Mao C, MacNeil T, Kalyani R, Smith T, Weinberg D, Tota MR, and Van der Ploeg LH (1997) ART (protein product of agouti-related transcript) as an antagonist of MC-3 and MC-4 receptors, *Biochem. Biophys. Res. Commun.* 3: 629–631.

Fuhlendorff J, Johansen NL, Melberg SG, Thogersen H, and Schwartz TW (1990a) The antiparallel pancreatic polypeptide fold in the binding of neuropeptide Y to Y1 and Y2 receptors. *J. Biol. Chem.* 265: 11706-11712.

Fuhlendorff J, Gether U, Aakerlund L, Langeland-Johansen N, Thogersen H, Melberg SG, Olsen UB, Thastup O, and Schwarz TW (1990b) [Leu31, Pro34] neuropeptide Y: a selective Y1 receptor agonist. *Proc. Natl. Acad. Sci. USA* 87: 182-186.

Gamber KM, Macarthur H, and Westfall TC (2005) Cannabinoids augment the release of neuropeptide Y in the rat hypothalamus. *Neuropharmacology* 5: 646-652.

Gardiner JV, Kong WM, Ward H, Murphy KG, Dhillo WS, and Bloom SR (2005) AAV mediated expression of anti-sense neuropeptide Y cRNA in the arcuate nucleus of rats results in decreased weight gain and food intake. *Biochem. Biopys. Res. Commun.* 4: 1088-1093.

Gehlert DR, Beavers LS, Johnson D, Gackenheimer SL, Schober DA, and Gadski RA (1996) Expression cloning of a human brain neuropeptide Y Y2 receptor. *Mol. Pharmacol.* 2: 224-228.

Gehlert DR, Schober DA, Gackenheimer SL, Beavers L, Gadski R, Lundell I, and Larhammar D (1997) [125I]Leu31, Pro34-PYY is a high affinity radioligand for rat PP1/Y4 and Y1 receptors: evidence for heterogeneity in pancreatic polypeptide receptors. *Peptides.* 18:397-401.

Gerald C, Walker MW, Vaysse PJ, He C, Branchek TA, and Weinshank RL (1995) Expression cloning and pharmacological characterization of a human hipoccampal neuropeptide Y/peptide YY Y2 receptor subtype. *J. Biol. Chem.* 45: 26758-26761.

Ghamari-Langroudi M, Srisai d, and Cone RD (2011) Multinodal regulation of the arcuate/paraventricular nucleus circuit by leptin. *Proc. Natl. Acad. Sci. USA* 1:355-360.

Ghersi G, Chen W, Lee EW, and Zukowska Z (2001) Critical role of dipeptidyl peptidase IV in neuropeptide Y-mediated endothelial cell migration in response to wounding. *Peptides* 22: 453-458.

Glover I, Haneef I, Pitts J, Woods S, Moss D, Tickle I, and Bundell T. (1983) Conformational flexibility in a small globular hormone: X-ray analysis of avian pancreatic polypeptide at 0.98-A resolution. *Biopolymers* 22: 293-304.

Glover ID, Barlow DJ, Pitts JE, Wood SP, Tickle IJ, and Blindell TL (1984) Conformational studies on the pancreatic polypeptide hormone family. *Eur. J. Biochem.* 142: 379-385.

Grouzmann E, and Brakch N. (2005) NPY processing in neuronal and non-neuronal tissues by proconvertases. In The NPY family of peptides in immune disorders, inflammation, angiogenesis, and cancer (Zukowska Z, and Feuerstein GZ eds) pp63-74 Birkhauser Verlag, Basel/Boston/Berlin.

Grove KL, Brogan RS, and Smith MS (2001) Novel expression of neuropeptide Y (NPY) mRNA in hypothalamic regions during development: region-specific effects of maternal deprivation on NPY and agouti-related protein mRNA. *Endocrinology* 11: 4771-4776.

Grove KL, Allen S, Grayson BE, and Smith MS (2003a) Postnatal development of the hypothalamic neuropeptide Y system. *Neurosci.* 2: 393-406.

Grove KL, and Smith MS (2003b) Ontogeny of the hypothalamic neuropeptide Y system. *Physiol Behav.* 1: 47-63.

Hahn TM, Breininger JF, Baskin DG, and Schwartz MW (1998) Coexpression of Agrp and NPY in fasting-activated hypothalamic neurons. *Nat. Neurosci.* 1: 271-272.

Halaas JL, Gaijiwala KS, Maffei M, Cohen SL, Chait BT, Rabinowitz D, Lallone RL, Burley SK, and Friedman JM (1995) Weight-reducing effects of the plasma protein encoded by the obese gene. *Science* 5223: 543-546.

Han S, Chen X, Cox B, Yang CL, Wu YM, Naes L, and Westfall T (1998) Role of neuropeptide Y in cold stress-induced hypertension. *Peptides* 19: 351-358.

Herzog H, Hort YJ, Ball HJ, Hayes G, Shine J, and Selbie LA (1992) Cloned human neuropeptide Y receptor couples to two different second messenger systems. *Proc. Natl. Acad. Sci. USA* 13: 5794-5798.

Herzog H, Darby K, Ball H, Hort Y, Beck-Sickinger A, and Shine J (1997) Overlapping gene structure of the human neuropeptide Y receptor subtypes Y1 and Y5 suggests coordinate transcriptional regulation. *Genomics.* 41:315-9.

Higuchi H, Yang HY, and Sabol SL (1988) Rat neuropeptide Y precursor gene expression, mRNA, structure, tissue distribution, and regulation by glucocorticoids, cyclic AMP, and phorbol ester. *J. Biol. Chem.* 263: 6288-6295.

Hipskind PA, Lobb KL, Nixon JA, Britton TC, Bruns RF, Catlow J, Dieckman-McGinty DK, Gackenheimer SL, Gitter BD, Iyengar S, Schober DA, Simmons RM, Swanson S, Zarrinmayeh H, Zimmerman DM, and Gehlert DR (1997) Potent and selective 1,2,3-trisubstituted indole NPY Y1 antagonists. *J. Med. Chem.* 23: 3712-3714.

Holliday ND, Michel MC, and Cox HM (2004) NPY receptor subtypes and their signal transduction, in Neuropeptide Y and related peptides (Michel MC eds) pp 46-67, *Springer.*

Hoyda TD, Smith PM, and Ferguson AV (2009) Adiponectin acts in the nucleus of the solitary tract to decrease blood pressure by modulating the excitability of neuropeptide Y neurons. *Brain Res.,* 76-84.

Hu Y, Bloomquist BT, Cornfield LJ, DeCarr LB, Flores-Riveros JR, Friedman L, Jiang P, Lewis-Higgins L, Sadlowski Y, Shaefer J, Velazquez N, and McCaleb ML (1996) Identification of a novel hypothalamic neuropeptide Y receptor associated with feeding behavior. *J. Biol. Chem.* 42: 26315-26319.

Huang XF, Han M, and Storlien LH (2003) The level of NPY receptor mRNA expression in diet-induced obese and resistant mice. *Mol. Brain Res.* 1: 21-28.

Hung CC, Pirie F, Luan J, Lank E, Motala A, Yeo GS, Keogh JM, Wareham NJ, O'Rahilly S, and Farooqi IS (2004) Studies of the peptide YY and neuropeptide Y2 receptor genes in relation to human obesity and obesity-related traits. *Diabetes* 53: 2462-2466.

Hunt SC, Hasstedt SJ, Xin Y, Dalley BK, Milash BA, Yakobson E, Gress RE, Davidson LE, and Adams TD (2011) Polymorphisms in the NPY2R gene show significant associations with BMI that are additive to FTO, MC4R, and NPFFR2 gene effects. *Obesity* 19: 2241-2247.

Ikeda H, West DB, Pustek JJ, Figlewicz DP, Greenwood MR, Porte Jr D, and Woods SC (1986) Intraventricular insulin reduces food intake and body weight of lean but not obese Zucker rats. *Appetite* 4: 381-386.

Inui A (1999) Neuropeptide Y feeding receptors: are multiple subtypes involved? *Trends Pharmacol. Sci.* 20: 43-46.

Ishihara A, Tanaka T, Kanatani A, Fukami T, Ihara M, and Fukuroda T (1998) A potent neuropeptide Y antagonist, 1229U91, suppressed spontaneous food intake in Zucker fatty rats. *Am. J. Physiol.* 2: R1500-1504.

Jacques D, Dumont Y, Fournier A, and Quirion R (1997) Characterization of neuropeptide Y receptor subtypes in the normal human brain, including the hypothalamus. *Neuroscience* 79: 129-148.

Jain MR, Horvath TL, Kalra PS, and Kalra SP (2000) Evidence that NPY Y1 receptors are involved in stimulation of feeding by orexins (hypocretins) in sated rats. *Regul. Pept.* 1-3: 19-24.

Jamshidi N, and Taylor DA (2001) Anandamide administration into ventromedial hypothalamus stimulates appetite in rats. *Br. J. Pharmacol.* 6: 1151-1154.

Kaga T, Inui A, Okita M, Asakawa A, Ueno N, Kasuga M, Fujimiya M, Nishimura N, Dobashi R, Morimoto Y, Liu IM, and Cheng JT (2001) Modest overexpression of neuropeptide Y in the brain leads to obesity after high sucrose feeding. *Diabetes* 5: 1206-1210.

Kamegai J, Tamura H, Shimizu T, Ishii S, Sugihara H, and Wakabayashi I (2001) Chronic central infusion of ghrelin increases hypothalamic neuropeptide Y and agouti-related protein mRNA levels and body weight in rats. *Diabetes* 11: 2438-2443.

Kanatani A, Ishihara A, Asahi S, Tanaka T, Ozaki S, and Ihara M (1996) Potent neuropeptide Y Y1 receptor antagonist, 1229U91: blockade of neuropeptide Y-induced and physiological food intake. *Endocrinology* 137: 3177–3182.

Kanatani A, Ito J, Ishihara A, Iwaasa H, Fukuroda T, Fukami T, MacNeil DJ, Van der Ploeg LH, and Ihara M (1998) NPY-induced feeding involves the action of a Y1-like receptor in rodents. *Regul Pept.* 75–76: 409–415.

Kanatani A, Kanno T, Ishihara M, Hata M, Sakuraba A, Tanaka T, Tsuchiya Y, Mase T, Fukuroda T, Fukami T, and Ihara M (1999) The novel neuropeptide Y Y (1) receptor antagonist J-104870: a potent feeding suppressant with oral bioavailability. *Biochem. Biophys. Res. Commun.* 1: 88-91.

Kanatani A, Mashiko S, Murai N, Sugimoto N, Ito J, Fukuroda T, Fukami T, Morin N, MacNei DJ, Van der Ploeg LH, Saga Y, Nishimura S, and Ihara M (2000) Role of the Y1 receptor in the regulation of neuropeptide Y-mediated feeding: comparison of wild-type, Y1 receptor-deficient, and Y5 receptor-deficient mice. *Endocrinology* 3: 1011-1016.

Kanatani A, Hata M, Mashiko S, Ishihara A, Okamoto O, Haga Y, Ohe T, Kanno T, Murai N, Ishii Y, Fukuroda T, Fukami T, and Ihara M (2001) A typical Y1 receptor regulates feeding behavior: effects of a potent and selective Y1 antagonist, J-115814. *Mol. Pharmacol.* 3:501-505.

Kannoa T, Kanatani A, Keen SL, Arai-Otsuki S, Haga Y, Iwama T, Ishihara A, Sakuraba A, Iwaasa H, Hirose M, Morishima H, Fukami T, and Ihara M (2001) Different binding sites for the neuropeptide Y Y1 antagonists 1229U91 and J-104870 on human Y1 receptors. *Peptides* 22: 405–413.

Kask A, Rago L, and Harro J (1998) Evidence for involvement of neuropeptide Y receptors in the regulation of food intake: studies with Y1-selective antagonist BIBP3226. *Br. J. Pharmacol.* 124: 1507–1515.

Karvonen MK, Pesonen U, Koulu M, Niskanen L, Laakso M, Rissanen A, Dekker M, Hart LM, Valve R, Uusitupa MI (1998) Association of a leucine(7)-to-proline(7) polymorphism in the signal peptide of neuropeptide Y with high serum cholesterol and LDL cholesterol levels. *Nat. Med.* 4: 1434-1437

Karvonen MK, Koulu M, Pesonen U, Uusitupa U, Tammi A, Viikari Simell O, and Rönnemaa T (2000) Leucine 7 to proline 7 polymorphism in the preproneuropeptide Y is associated with birth weight and serum triglyceride concentration in preschool aged children. *J. Clin. Endocrinol. Metab.* 85: 1455-1460.

Kawamura K, Smith TL, Zhou Q, and Kummerow FA (1991) Neuropeptide Y stimulates prostacyclin production in porcine vascular endothelial cells. *Biochem. Biophys. Res. Commun.* 179: 309-313.

Keen-Rhinenhart E, and Bartness TJ (2007) NPY Y1 receptor is involved in ghrelin- and fasting-induced increases in foraging, food hoarding, and food intake. *Am. J. Physiol. Regul. Integr. Comp. Physiol.* 4: R1728-1737.

Kerr NC, Holmes FE, and Wynick D (2000) Galanin-like peptide (GALP) is expressed in rat hypothalamus and pituitary, but not in DRG, *Neuroreport* 17: 3909–3913.

Kirkham TC, Williams CM, Fezza F, and Di Marzo V (2002) Endocannabinoid levels in rat limbic forebrain and hypothalamus in relation to fasting, feeding and satiation: stimulation of eating by 2-arachidonoyl glycerol. *Br. J. Pharmacol.* 4: 550-557

Kohno D, Gao HZ, Muroya S, Kikuyama S, and Yada T (2003) Ghrelin directly interacts with neuropeptide-Y-containing neurons in the rat arcuate nucleus: Ca^{2+} siganaling via protein kinase A and N-type channel-dependent mechanisms and cross-talk with leptin and orexin. *Diabetes* 4: 948-956.

Kohno D, Nakata M, Maekawa F, Fujiwara K, Maejima Y, Kuramochi M, Shimazaki T, Okano H, Onaka T, and Yada T (2007) Leptin suppresses ghrelin-induced activation of neuropeptide Y neurons in the arcuate nucleus via phosphatidylinositol 3-kinase and phosphodiesterase 3-mediated pathway. *Endocrinology* 5: 2251-2263.

Kohno D, Sone H, Minokoshi Y, and Yada T (2008) Ghrelin raises $[Ca^{2+}]i$ via AMPK in hypothalamic arcuate nucleus NPY neurons. *Biochem. Biophys. Res. Commun.* 2: 388-392.

Korner J, Savontaus E, Cha Jr SC, Leibel RL, and Wardlaw SL (2001) Leptin regulation of Agrp and Npy mRNA in the rat hypothalamus. *J. Neuroendocrinol.* 11: 959-966.

Kubota N, Yano W, Kubota T, Yamauchi T, Itoh S, Kumagai H, Kozono H, Takamoto I, Okamoto S, Shiuchi T, Suzuki R, Satoh H, Tsuchida A, Moroi M, Sugi K, Noda T, Ebinuma H, Ueta Y, Kondo T, Araki E, Ezaki O, Nagai R, Tobe K, Terauchi Y, Ueki K, Minokoshi Y, and Kadowaki T (2007) Adiponectin stimulates AMP-activated protein kinase in the hypothalamus and increases food intake. *Cell Metab.* 1: 55-68.

Kuo LE, Kitlinska JB, Tilan JU, Li L, Baker SB, Johnson MD, Lee EW, Burnett MS, Fricke ST, Kvetnansky R, Herzog H, and Zukowska Z (2007) Neuropeptide Y acts directly in the periphery on fat tissue and mediates stress-induced obesity and metabolic syndrome. *Nat. Med.* 13: 803-811.

Kuramochi M, Onaka T, Kohno D, Kato S, and Yada T (2006) Galanin-like peptide stimulates food intake via activation of neuropeptide Y neurons in the hypothalamic dorsomedial nucleus of the rat. *Endocrinology* 4: 1744-1752.

Kushi A, Sasai H, Koizumi H, Takeda N, Yokoyama M, and Nakamura M (1998) Obesity and mild hyperinsulinemia found in neuropeptide Y-Y1 receptor-deficient mice. *Proc. Natl. Acad. Sci. USA* 26: 15659-15664.

Larhammar D, Ericsson A, and Persson H (1987) Structure and expression of the rat neuropeptide Y gene. *Proc. Natl. Acad. Sci. USA* 84: 2068-2072.

Larhammar D, Blomqvist AG, Jazin YF, Yoo H, and Wahlested C (1992) Cloning and functional expression of a human neuropeptide Y/peptide YY receptor of the Y1 type. *J. Biol. Chem.* 267: 10935-10938.

Larhammar D, and Salaneck E (2004) Molecular evolution of NPY receptor subtypes. *Neuropeptides* 4: 141-151.

Larsen PJ, Tang-Christensen M, Stidsen CE, Madsen K, Smith MS, and Cameron JL (1999) Activation of central neuropeptide Y Y1 receptors potently stimulates food intake in male rhesus monkeys. *J. Clin. Endocrinol. Metab.* 84: 3781–3791.

Lavebratt C, Alpman A, Persson B, Arner P, and Hoffstedt J (2006) Common neuropeptide Y2 receptor gene variant is protective against obesity among Swedish men. *Int. J. Obes.* 30: 453-459.

Lecklin A, Lundell I, Salmela S, Manisto PT, Beck-Sickinger AG, and Larhammar D (2003) Agonists of neuropeptide Y receptors Y1 and Y5 stimulate different phases of feeding in guinea pigs. *Br. J. Pharmacol.* 8: 1433-1440.

Lee EW, Grant DS, Movafagh S, and Zukowska Z (2003a) Impaired angiogenesis in neuropeptide Y (NPY)-Y2 receptor knockout mice. *Peptides* 24: 99-106.

Lee EW, Michalkiewicz M, Kitlinska J, Kalezic I, Switalska H, Yoo P, Sangkharat A, Ji H, Li L, Michalkiewicz T, Ljubisavljevic M, Johansson H, Grant DS, and Zukowska Z (2003b) Neuropeptide Y induces ischemic angiogenesis and restores function of ischemic skeletal muscles. *J. Clin. Invest.* 111: 1853–1862.

Li JY, Finniss S, Yang YK, Zeng Q, Qu SY, Barsh G, Dickinson C, and Ganz I (2000) Agouti-related protein-like immunoreactivity: characterization of release from hypothalamic tissue and presence in serum. *Endocrinology* 6: 1942-1950.

Li L, Lee EW, Ji H, and Zukowska H (2003) Neuropeptide Y-induced acceleration of post-angioplasty occulusion of rat carotid artery. Arterioscler *Thromb. Vasc. Biol.* 23: 1204-1210.

Lumb KJ, DeCarr LB, MilardoLF, Mays MR, Buckholz TM, Fisk SE, Pellegrino CM, Ortiz AA, and Mahle CD (2007) Novel selective neuropeptide Y2 receptor PEGylated peptide agonists reduce food intake and body weight in mice. *J. Med. Chem.* 9: 2264-2268.

Lundberg JM, and Modin A (1995) Inhibition of sympathetic vasoconstriction in pigs in vivo by the neuropeptide Y-Y1 receptor antagonist BIBP 3226. *Br. J. Pharmacol.* 116: 2971-2982.

Lutz TA, Del Prete E, and Scharrer E (1994) Reduction of food intake in rats by intraperitoneal injection of low doses of amylin. *Physiol. Behav.* 5: 891-895.

Lutz TA, Rossi R, Althaus J, Del Prete E, and Scharrer E (1998) Amylin reduces food intake more potently than calcitonin gene-related peptide (CGRP) when injected into the lateral brain ventricle in rats. *Peptides* 9: 1533-1540.

Ma L, Tataranni PA, Hanson RL, Infante AM, Kobes S, Bogardus C, and Baier LJ (2005) Variations in peptide YY and Y2 receptor genes are associated with severe obesity in Pima Indian men. *Diabetes* 54: 1598-1602.

Malmstrom RE, and Lundberg JM (1995a) Neuropeptide Y accounts for sympathetic vasoconstriction in guinea-pig vena cava: evidence using BIBP 3226 and 3435. *Eur. J. Pharmacol.* 294: 661-668.

Malmstrom RE, and Lundberg JM (1995b) Endogenous NPY acting on the Y1 receptor accounts for the long-lasting part of the sympathetic contraction in guinea-pig vena cava: evidence using SR 120107A. *Acta Physiol. Scand.* 155: 329-330.

Malmstrom RE (2001) Vascular pharmacology of BIIE0246, the first selective non-peptide neuropeptide Y Y2 receptor antagonist, in vivo. *Br. J. Pharmacol.* 133: 1073-1080.

Malmstrom RE (2002) Pharmacology of neuropeptide Y receptor antagonists. Focus on cardiovascular functions. *Eur. J. Pharmacol.* 447: 11-30.

Marsh DJ, Hollopeter G, Kafer KE, and Palmiter RD (1998) Role of the Y5 neuropeptide Y receptor in feeding and obesity. *Nat. Med.* 6: 718-721.

McCrea K, Wisialowski T, Caberele C, Church B, Beck-Sickinger A, Kraegen E, and Herzog H (2000) 2-36[K4,RYYSA(19-23)]PP a novel Y5-receptor preferring ligand with strong stimulatory effect on food intake. *Regul. Pept.* 1-3: 47-58.

Melander T, Hokfelt T, and Rokaeus A (1986) Distribution of galanin like immunoreactivity in the rat central nervous system. *J. Comp. Neurol.* 4: 475-517.

Michel MC, Beck-Sickinger A, Cox H, Doods HN, Herzog H, Larhammar D, et al. (1998) XVI. International Union of Pharmacology recommendations for the nomenclature of neuropeptide Y, peptide YY, and pancreatic polypeptide receptors. *Pharmacol. Rev.* 50: 143-150.

Montague CT, Farooqi IS, Whitehead JP, Soos MA, Rau H, Wareham NJ, Sewter CP, Digby JE, Mohammed SN, Hurst JA, Cheetham CH, Earley AR, Barnet AH, Prins JB, and O'Rahilly S (1997) Congenital leptin deficiency is associated with severe early-onset obesity in humans. *Nature* 6636: 903-908.

Moreno G, Perello M, Gaillard RC, and Spinedi E (2005) Orexin a stimulates hypothalamic-pituitary-adrenal (HPA) axis function, but not food intake, in the absence of full hypothalamic NPY-ergic activity, *Endocrine* 2: 99–106.

Morgan DG, Small CJ, Abusnana S, Turton M, Gunn I, Heath M, Rossi M, Goldstone AP, O'Shea D, Meeran K, Ghatei M, Smith DM, and Bloom S

(1998) The NPY Y1 receptor antagonist BIBP 3226 blocks NPY induced feeding via a non-specific mechanism. *Regul. Pept.* 75–76: 377–382.

Morley JE, and Flood JF (1991) Amylin decreases food intake in mice. *Peptides* 4: 865-869.

Moser C, Bernhardt G, Michel J, Schwarz H, and Buschauer A (2000) Cloning and functional expression of the hNPY Y5 receptor in human endometrial cancer (HEC-1B) cells. *Can. J. Physiol. Pharmacol.* 78:134-142

Mullins D, Kirby D, Hwa J, Guzzi M, Rivier J, and Parker E (2001) Identification of potent and selective neuropeptide Y Y(1) receptor agonists with orexigenic activity in vivo. *Mol. Pharmacol.* 3: 534-540.

Nakamura NM, Sakanaka C, Aoki Y, Ogasawara H, Tsuji T, Kodama H, Matsumoto T, Shimizu T, and Noma M (1995) Identification of two isoforms of mouse neuropeptide Y-Y1 receptor generated by alternative splicing. Isolation, genomic structure, and functional expression of the receptors. *J. Biol. Chem.* 50: 30102-30110.

Nakazato M, Murakami N, Date Y, Kojima M, Matsuo H, Kangawa K, and Matsukura S (2001) A role for ghrelin in the central regulation of feeding. *Nature* 6817: 194-198.

Naveilhan P, Hassani H, Canals JM, Ekstrand AJ, LarefalkA, Chhajlani V, Arenas E, Gedda K, Svensson L, Thoren P, and Ernfors P (1999) Normal feeding behavior, body weight and leptin response require the neuropeptide Y Y2 receptor. *Nat. Med.* 10: 1188- 1193.

Niskanen L, Karvonen MK, Valve R, Koulu M, Pesonen U, Mercuri M et al. (2000) Leucine 7 to proline 7 polymorphism in the neuropeptide Y gene is associated with enhanced carotid atherosclerosis in elderly patients with type 2 diabetes and control subjects. *J. Clin. Endocrinol. Metab.* 85: 2266-2269.

Noll T, Hampel A, and Piper HM (1996) Neuropeptide Y reduces macromolecule permeability of coronary endothelial monolayers. *Am. J. Physiol.* 271: H1878-1883.

Obici S, Feng Z, Karkanias G, Baskin DG, and Rossetti L (2002) Decreasing hypothalamic insulin receptors causes hyperphagia and insulin resistance in rats. *Nat. Neurosci.* 6: 566-572.

Ollmann MM, Wilson BD, Yang YK, Kerns JA, Chen Y, Ganz I, and Barsh GS (1997) Antagonism of central melanocortin receptors in vitro and in vivo by agouti-related protein. *Science* 278: 135-138.

Ortiz AA, MilardoLF, DeCarr LB, Buckholz TM, Mays MR, Claus TH, Livingston JN, Mahle CD, and Lumb KJ (2007) A novel long-acting selective neuropeptide Y2 receptor polyethylene glycol-conjugated peptide agonist reduces food intake and body weight and improves glucose metabolism in rodents. *J. Pharmacol. Exp. Ther.* 2: 692-700.

Pankajakshan D, Jia G, Pipinos I, Tyndall SH, and Agrawal DK (2011) Neuropeptide Y receptors in carotid plaques of symptomatic and asymptomatic patients: effect of inflammatory cytokines. *Exp. Mol. Pathology* 90: 280-286.

Parker EM, Balasubramaniam A, Guzzi M, Mullins DE, Salisbury BG, Sheriff S, Witten MB, and Hwa JJ (2000) D-Trp(34) neuropeptide Y is a potent and selective neuropeptide Y Y(5) receptor agonist with dramatic effects on food intake. *Peptides* 3: 393-399.

Parker SL, Parker MS, Sah R, Balasubramaniam A, and Sallee FR (2008a) Pertussis toxin induces parallel loss of neuropeptide Y Y1 receptor dimers and Gi alpha subunit function in CHO cells. *Eur. J. Pharmacol.* 579:13-25.

Parker SL and Balasubramaniam A (2008b) Neuropeptide Y Y2 receptor in health and disease. *Br. J. Pharmacol.* 153:420-31.

Parrado C, Diaz-Cabiale Z, Galcia-Coronel M, Agnati LF, Covenas R, Fuxe K, and Narvaez JA (2007) Region specific galanin receptor/neuropeptide Y Y1 receptor interactions in the tel- and diencephalon of the rat. Relevance for food consumption. *Neuropharmaology* 2: 684-692.

Pedrazzini T, Seydoux J, Kunstner P, Aubert JF, Grouzmann E, Beermann F, and Brunner HR (1998) Cardiovascular response, feeding behavior and locomoter activity in mice lacking the NPY Y1 receptor. *Nat. Med.* 6: 722-726.

Pelleymounter MA, Cullen MJ, Baker MB, Hecht R, Winters D, Boone T, and Collins F (1995) Effects of the obese gene product on body weight regulation in ob/ob mice. *Science* 5223: 540-543.

Pich EM, Messori B, Zoi M, Ferraguti F, Marrama P, Biagini G, Fuxe K, and Agnati LF (1992) Feeding and drinking responses to neuropeptide Y injections in the paraventricular hypothalamic nucleus of aged rats. *Brain Res.* 2: 265-271.

Pittner RA, Moore CX, Bhavsar SP, Gedulin BR, Smith PA, Jodka CM, Parkes DG, Paterniti JR, Srivasatava VP, and Young AA (2004) Effect of PYY [3-36] in rodent models of diabetes and obesity. *Int. J. Obest. Relat. Metab. Disord.* 8: 963-971.

Potter EK, and McCloskey MJ (1992) [Leu31, Pro34] NPY, a selective functional postjunctional agonist at neuropeptide-Y receptors in anaesthetized rats. *Neurosci. Lett.* 134: 183-186.

Pralong FP, Gonzales C, Voirol MJ, Palmiter RD, Brunner HR, Gailard RC, Seydoux J, and Pedrazzini T (2002) The neuropeptide Y Y1 receptor regulates leptin-mediated control of energy homeostasis and reproductive functions. *FASEB J.* 7: 712-714.

Pronchuk N, Beck-Sickinger AG, and Colmers WF (2002) Multiple NPY recptors inhibit GABA (A) synaptic responses of rat medial parvocellular effector neurons in the hypothalamic paraventiricular nucleus. *Endocrinology* 2: 535-543.

Qureshi NU, Dayao EK, Shirali S, Zukowska-Grojec Z, and Hauser GJ (1998) Endogenous neuropeptide Y mediates vasoconstriction during endotoxic and hemorrhagic shock. *Regul. Pept.* 75-76: 215-220.

Roche C, Boutin P, Dina C, Gyapay G, Basdevant A, Harger J, Guy-Grand B, Clement K, and Froguel P (1997) Genetic studies of neuropeptide Y and neuropeptide Y Y1 and Y5 regions in morbid obesity. *Diabetologia* 40: 671-675.

Rodriguez M, Audinot V, Dromaint S, Macia C, Lamamy V, Beauverger P, Rique H, Imbert J, Nicolas JP, Boutin JA, and Galizzi JP (2003) Molecular identification of the long isoform of the human neuropeptide Y Y5 receptor and pharmacological comparison with the short Y5 receptor isoform. *Biochem. J. Pt.* 3: 667-673.

Rose PM, Fernandes P, Lynch JS, Fraizier ST, Fisher SM, Kodukula K, Kienzle B, and Seethala R (1995) Cloning and functional expression of a cDNA encoding a human type 2 neuropeptide Y receptor. *J. Biol. Chem.* 39: 22661-22664.

Rudolf K, Eberlein W, Engel W, Wieland HA, Willim KD, Entzeroth M, Wienen W, Beck-Sickinger AG, and Doods HN (1994) The first highly potent and selective non-peptide neuropeptide Y Y1 receptor antagonist: BIBP 3226. *Eur. J. Pharmacol.* 271: R11–R13.

Sainsbury A, Shwarzer C, Couzens M, Fetisov S, Furtinger S, Jenkins A, Cox HM, Sperk G, Hokfelt T, and Herzog H (2002) Important role of hypothalamic Y2 receptors in body weight regulation revealed in conditional knockout mice. *Proc. Natl. Acad. Sci. USA* 13: 8938-8943.

Sainsbury A, Bergen HT, Boey D, Bamming D, Conney GJ, Lin S, Couzens M, Stroth N, Lee NJ, Lindner D, Singewald N, Karl T, Duffy L, Enriquez R, Slack K, Sperk G, and Herzog H (2006) Y2Y4 receptor double knockout protects against obesity due to a high-fat diet or Y1 receptor deficiency in mice. *Diabetes* 1: 19-26.

Sautel M, Martinez R, Munoz M, Peitsch MC, Beck-Sickinger AG, and Walker P. (1995) Role of a hydrophobic pocket of the human Y1 neuropeptide Y receptor in ligand binding. *Mol. Cell Endocrinol.* 112: 215-222.

Sautel M, Rudolf K, Wittneben H, Herzog H, Martinetz R, Munoz M, Eberlein W, Engel W, Walker P, and Beck-Sickinger AG (1996) Neuropeptide Y and the nonpeptide antagonist BIBP3226 share an overlapping binding site at the human Y1 receptor. *Mol. Pharmacol.* 50: 285-292.

Scott V, Kimura N, Stark JA, and Luckman SM (2005) Intravenous peptide YY3-36 and Y2 receptor antagonism in the rat: effect on feeding behavior. *J. Neuroendocrinol.* 7: 452-457.

Segal-Lieberman G, Trombly DJ, Juthani V, Wang X, and Maratos-Flier E (2003) NPY ablation in C57BL/6 mice leads to mild obesity and to an impaired refeeding response to fasting. *Am. J. Physiol. Endocrinol. Metab.* 6: E131-139.

Shimada K, Ohno Y, Okamatsu-Ogura Y, Suzuki M, Kamikawa A, Terao A, and Kimura K (2012) Neuropeptide Y activates phospholylation of ERK and STAT3 in stromal vascular cells from brown adipose tissue, but fails to affect thermogenic function of brown adipocytes. *Peptides* 2: 336-342.

Shionogi and Co. (2009) Shionogi announces positive top-line efficacy results from year-long studies of velneperit, a novel NPY Y5 receptor antagonist being investigated for the treatment of obesity, *http://www.shionogi.co.jp /ir_en/news/detail/e_090217-2.pdf#search='Shionogi%20velneperit'.*

Shoblock JR, Welty N, Nepomuceno D, Lord B, Aluisio L, Fraser I, Motley ST, Sutton SW, Morton K, Galici R, Atack JR, Dvorak L, Swanson DM, Carthers NI, Dvorak C, Lovenberg TW, and Bonaventure P (2010) In vitro and in vivo characterization of JNJ-31020028(N-(4-(4-[2-(diethylamino)-2-oxo1-phenylethyl]piperazin-1-yl)-3-fluorophenyl)-2-pyridine-3-ylbenzamide), a selective brain penetrant small molecule anatagonist of the neuropeptide Y(2) receptor. *Psychopharmacology* (Berl) 2: 265-277.

Silva AP, Cavadas C, and Grouzmann E (2002) Neuropeptide Y and its receptors as potential therapeutic drug targets. *Clin. Chim. Acta* 326: 3-25.

Siddiq A, Gueorguiev M, Samson C, Hereberg S, Heude B, Levy-Marchal C, Jouret B, Weill J, Meyre D, Walley A, and Froguel P (2007) Single nucleotide polymorphisms in the neuropeptide Y2 receptor (NPY2R) gene and association with severe obesity in French white subjects. *Diabetologia* 50: 574-584.

Sjodin P, Holmberg SK, Akerberg H, Berglund MM, Mohell N, and Larhammar D (2006) Re-evaluationof receptor-ligand interactions of the human neuropeptide Y receptor Y1: a site-directed mutagenesis study. *Biochem. J.* 393: 161-169.

Skofitsch G, and Jacobowitz DM (1985) Immunohistochemical mapping of galanin-like neurons in the rat central nervous system. *Peptides* 3: 509-546.

Stanley BG, Chin AS, and Leibowitz SF (1985) Feeding and drinking elicited by central injection of neuropeptide Y: evidence for a hypothalamic site(s) of action. *Brain Res. Bull.* 6: 521-524.

Stanley BG, Magdalin W, Seirafi A, Nguyen MM, and Leibowitz SF (1992) Evidence for neuropeptide Y mediation of eating produced by food deprivation and for a variant of the Y1 receptor mediating this peptide's effect. *Peptides* 3: 581-587.

Statnick MA, Schober DA, Gackenheimer S, Johnnson D, Beavers L, Mayne NG, Burnett JP, Gadski R, and Gehlert DR (1998) Characterization of the neuropeptide Y5 receptor in the human hypothalamus: a lack of correlation between Y5 mRNA levels and binding cites. *Brain Res.* 1-2: 16-26.

Takahashi KA, and Cone RD (2005) Fasting induces a large, leptin-dependent increase in the intrinsic action potential frequency of orexigenic arcuate nucleus neuropeptide Y/Agouti-related protein neurons. *Endocrinology* 3: 1043-1047.

Takeuchi T, Gumucio DL, Yamada T, Meisler MH, Minth CD, and Dixon JE (1986) Genes encoding pancreatic polypeptide and neuropeptide Y are on human chromosomes 17 and 7. *J. Clin. Invest.* 77:1038-1041.

Takiguchi E, Fukano C, Kimura Y, Tanaka M, Tanida K, and Kaji H (2010) Variation in the 5'-flanking region of the neuropeptide Y2 receptor gene and metabolic parameters. *Metabolism* 59: 1591-1596.

Takiyyuddin MA, Brown MR, Dinh TQ, Cervenka JH, Braun RJ, Parmer RJ, et al. (1994) Sympatho-adrenal secretion in humans:factors governing catecholamine and storage vesicle peptide co-release. *J. Auton. Pharmacol.* 14: 187–92.

Tatemoto K, Carlquist M, and Mutt V (1982) Neuropeptide Y – a novel brain peptide with structural similarities to peptide YY and pancreatic polypeptide. *Nature* 5858:659-660.

Tiesjema B, la Fleur SE, Luijendijk MC, Brans MA, Lin EJ, During MJ, and Adan RA (2007a) Viral mediated neuropeptide Y expression in the rat paraventricular nucleus results in obesity in Obesity pp2424-2435, *Silver Spring 1.*

Tiesjema B, Adan RA, Luijendijk MC, Kalsbeek A, and la Fleur SE (2007b) Differential effects of recombinant adeno-associated virus-mediated neuropeptide Y overexpression in the hypothalamic paraventricular nucleus and lateral hypothalamus on feeding behavior. *J. Neurosci.* 51: 14139-14146.

Tiesjema B, la Fleur SE, Luijendijk MC, and Adan RA (2009) Sustained NPY overexpression in the PVN results in obesity via temporally increasing food intake, in Obesity pp1448-1450, *Silver Spring*.

Torekov SS, Larsen LH, Andersen G, Albrechtsen A, Glumer C, Borsh-Johnsen K, Jorgensen T, Hansen T, and Pedersen O (2006) Variants in the 5' region of the neuropeptide Y receptor Y2 gene (NPY2R) are associated with obesity in 5,971 white subjects. *Diabetologia* 49: 2653-2658.

Toshinai K, Date Y, Murakami N, Shimada M, Mondal MS, Shimbara T, Guan JL, Wang QP, Funahashi H, Sakurai T, Shioda S, Matsukura S, Kangawa K, and Nakazato M (2003) Ghrelin-induced food intake is mediated via the orexin pathway. *Endocrinology* 4: 1506-1512.

Tschop M, Simley DL, and Heiman ML (2000) Ghrelin induces adiposity in rodents. *Nature* 6806: 908-913.

Van den Pol AN, Acuna-Goycolea C, Clark KR, and Ghosh PK (2004) Physiological properties of hypothalamic MCH neurons identified with selective expression of reporter gene after recombinant virus infection. *Neuron* 4: 635-652.

Van den Top M, Lee K, Whyment AD, Blanks AM, and Spanswick D (2004) Orexigen-sensitive NPY/AgRP pacemaker neurons in the hypothalamic arcuate nucleus. *Nat. Neurosci.* 5: 493-494.

Van der Kolk N, Madison FN, Mohr M, Eberhard N, Kofler B, and Fraley GS (2010) Alarin stimulates food intake in male rats and LH secretion in castrated male rats. *Neuropeptides* 4: 333-340.

Verty AN, Boon WM, Mallet PE, McGregorI S, and Oldfield BJ (2009) Involvement of hypothalamic peptides in the anorectic action of the CB receptor antagonist rimonabant (SR 141716). *Eur. J. Neurosci.* 11: 2207-2216.

Vrang N, Madsen AN, Tang-Christensen M, Hansen G, and Larsen PJ (2006) PYY (3-36) reduces food intake and body weight and improves insulin sensitivity in rodent models of diet-induced obesity. *Am. J. Physiol. Regul. Integr. Comp. Physiol.* 2: R367-375.

Wahlestedt C, Edvinsson L, Ekblad E, and Hakanson R (1985) Neuropeptide Y potentiates noradrenaline-evoked vasoconstriction: mode of action. *J. Pharmacol. Exp. Ther.* 234: 735-741.

Wahlestedt C, and Hakanson R (1986a) Effects of neuropeptide Y (NPY) at the sympathetic neuroeffector junction. Can pre- and postjunctional receptors be distinguished? *Med. Biol.* 64: 85-88.

Wahlestedt C, Yanaihara N, and Hakanson R (1986b) Evidence for different pre- and post-junctional receptors for neuropeptide Y and related peptides. *Regul. Pept.* 3: 307-318.

Wang C, Yang N, Wu S, Liu L, Sun X, and Nie S (2007) Difference of NPY and its receptor gene expression between obesity and obesity-resistant rats in response to high-fat diet. *Horm. Metab. Res.* 4: 262-267.

Weinberg DH, SirinathsinghiiDJ, Tan CP, Shiao LL, Morin N, Rigby MR, Heavens RH, Rapoport DR, Bayne ML, Cascieri MA, Strader CD, Linemyer DL, and MNeil DJ (1996) Cloning and expression of a novel neuropeptide Y receptor, *J. Biol. Chem.* 28: 16435- 16438.

Wharton J, Gordon L, Byrne J, Herzog H, Selbie LA, Moore K, Sullivan MH, Elder MG, Moscoso G, and Taylor KM (1993) Expression of the human neuropeptide tyrosine Y1 receptor. *Proc. Natl. Acad. Sci. USA* 2: 687-691.

Widdowson PS, Upton R, Henderson L, Buckingham R, Wilson S, and Williams G (1997) Reciprocal regional changes in brain NPY receptor density during dietary restriction and dietary-induced obesity in the rat. *Brain Res.* 1-2: 1-10.

Wieland HA, Engel W, Eberlein W, Rudolf K, and Doods HN (1998) Subtype selectivity of the novel nonpeptide neuropeptide Y Y1 receptor antagonist BIBO3304 and its effect on feeding in rodents. *Br. J. Pharmacol.* 3: 549-555.

Williams AG, Hargreaves AC, Gunn-Moore FJ, and Tavare JM (1998) Stimulation of neuropeptide Y gene expression by brain-derived neurotrophic factor requires both the phospholipase C gamma and Shc binding sites on its receptor, TrkB. *Biochem. J.* 333: 505-509.

Wood SP, Pitts JE, Blundell TL, Ticke IJ, and Jenkins JA (1977) Purification, crystallization and preliminary X-ray studies on avian pancreatic polypeptide. *Eur. J. Biochem.* 78: 119-126.

Woods SC, Lotter EC, McKay LD, and Porte Jr D (1979) Chronic intracerebroventricular infusion of insulin reduces food intake and body weight of baboons. *Nature* 5738: 503-505.

Wren AM, Seal LJ, Cohen MA, Brynes AE, Frost GS, Murphy KG, Dhillo WS, Ghatei MA, and Bloom SR (2001a) Ghrelin enhances appetite and increases food intake in humans. *J. Clin. Endocrinol. Metab.* 12: 5992.

Wren AM, Small CJ, Abbott CR, Dhillo WS, Seal LJ, Cohen MJ, Batterham RL, Taheri S, Stanley SA, Ghatei MA, and Bloom SR (2001b) Ghrelin causes hyperphagia and obesity in rats. *Diabetes* 11: 2540-2547.

Xin XG, and Huang XF (1998) Down-regulated NPY receptor subtype-5 mRNA expression in genetically obese mouse brain. *Neuroreport* 4: 737-741.

Yamanaka A, Kunii K, Nambu T, Tsujino N, Sakai A, Matsuzaki I, Miwa Y, Goto K, and Sakurai T (2000) Orexin-induced food intake involves neuropeptide Y pathway. *Brain Res.* 2: 404-409.

Yang L, Scott KA, Hyun J, Tamashiro KL, Tray N, Moran TH, and Bi S (2009) Role of dorsomedial hypothalamic neuropeptide Y in modulating food intake and energy balance. *J. Neurosci.* 1: 179-190.

Yasuda D, Okuno T, Yokomizo T, Hori T, Hirota N, Hashidate T, Miyano M, Shimizu T, and Nakamura M (2009) Helix 8 of leukotriene B4 type-2 receptor is required for the folding to pass the quality control in the endoplasmic reticulum. *FASEB J.* 23:1470-1481.

Zammaretti F, Panzica G, and Eva C (2001) Fasting, leptin treatment, and glucose administration differentially regulate Y (1) receptor gene expression in the hypothalamus of transgenic mice. *Endocrinology* 9: 3774-3782.

Zammaretti F, Panzica G, and Eva C (2007) Sex-dependent regulation of hypothalamic neuropeptide Y-Y1 receptor gene expression in moderate/high fat, high energy diet-fed mice. *J. Physiol.* Pt2: 445-454.

Zhang Y, Proenca R, Maffei M, Barone M, Leopold L, and Friedman JM (1994) Positional cloning of the mouse obese gene and its human homologue. *Nature* 6505: 425-432.

Zukowska-Grojec Z, Karwatowska-Prokopczuk E, Rose W, Rone J, Movafagh S, Ji H, Yeh Y, Chen WT, Kleinman HK, Grouzmann E, and Grant DS (1998) Neuropeptide Y: a novel angiogenic factor from the sympathetic nerves and endothelium. *Circ. Res.* 83: 187-195.

In: Neuropeptide Y
Editors: Steven L. Parker

ISBN: 978-1-62618-421-3
© 2013 Nova Science Publishers, Inc.

Chapter IV

Developmental Role of Neuropeptide Y in the Autonomic Ganglia

P. M. Masliukov[1] and A. D. Nozdrachev[2]
[1]Department of Normal Physiology, Yaroslavl State Medical
Academy, Yaroslavl, Russia
[2]Laboratory of Physiology of Reception, Pavlov Institute
of Physiology RAS, St. Petersburg, Russia

Abstract

Neuropeptide Y (NPY) is a 36-amino acid peptide widely distributed
in both the central and peripheral nervous systems. NPY and its receptors
play extremely diverse roles in the nervous system including regulation
blood pressure, circadian rhythms, feeding behavior, anxiety, vasoconstriction and gastrointestinal tract motility. In mammals, NPY has
been revealed in the majority of sympathetic ganglion neurons, in a high
number of neurons of parasympathetic cranial ganglia as well as of
intramural ganglia. NPY exerts its action via specific receptors,
designated as Y1 to Y6, of which Y1, Y2, Y4, Y5 and Y6 have been
cloned. All known NPY receptors belong to family A (rhodopsin-like) of
the large superfamily of G-protein-coupled heptahelical receptors.
Actions of NPY on peripheral target-organs are predominantly realized
through postsynaptic receptors Y1, Y3–Y5, and presynaptic Y2 receptors.

NPY is present in large dense-cored vesicles and is released at high-frequency stimulation. NPY affects not only vascular tone, frequency and strength of heart contractions, motorics and secretion of the gastrointestinal tract, but also has trophic effect and produces proliferation of cells of organs-targets, specifically of vessels, myocardium, and adipose tissue. During early postnatal development, the percentage of the NPY-containing neurons in many autonomic ganglia increases. In aged organisms, the proportion of NPY-immunopositive neurons decreases. This seems to be connected with the trophic NPY effect on target cells, as well as with regulation of their functional state.

Introduction

NPY and its receptors have been reported to be present in all major tissues of the body and implicated in numerous processes. NPY is widely distributed in both the central and peripheral nervous systems has been functionally related to regulation of blood pressure, circadian rhythms, feeding behavior, anxiety, memory processing, and cognition in the CNS, and to vaso-constriction and gastrointestinal tract motility in the PNS (Mutt et al., 1989; Allen and Koenig, 1992). Besides, NPY acts an important developmental factor. This peptide promotes proliferation and differentiation of a variety of cells. During embryogenesis, NPY expression results from a complex combination of regulatory cues. Our purpose in this chapter is to describe some recent additions to our knowledge about the development of NPY-containing autonomic neurons during the development.

Neuropeptide Y Structure and Synthesis

Neuropeptide Y (NPY) is a 36 -amino acid peptide including five tyrosine residues in each molecule and a C terminal amide structure (Tatemoto, 1982; Tatemoto et al., 1982). The strong conservation of the amino acid sequence of neuropeptide Y indicates that this peptide subserves evolutionarily old and important functions. The shark peptide displays only two differences by comparison to the sequence of the goldfish peptide and three as compared with the human peptide (Michel, 1991; McDermott et al., 1993). Human, rat, rabbit. and guinea pig forms of neuropeptide Y are identical, with methionine residue at position 17, which is readily oxidized (O'Hare et al., 1988). Porcine and bovine neuropeptide Y variants are identical to the mammalian forms described above except that the methionine residue in the latter at position 17 is replaced by a leucine residue which is not oxidized (McDermott et al., 1993).

Biologically active NPY is derived from a 97-amino acid precursor, preproneuropeptide Y. NPY is formed following four posttranslational enzymatic reactions (Higuchi et al., 1988), the first of which results in the 69 amino acid residue proNPY after the removal of the signal sequence by signal peptidase (Hodges et al., 2009). Subsequently, proNPY is broken down by prohormone convertase PC2 and/or PC1/3 at the paired basic site Lys38-Arg39, in turn, releasing the 30 amino acid C-flanking peptide of NPY and NPY(1–39) (Minth et al., 1984). A carboxypeptidase-like enzyme further processes NPY(1–39), resulting in NPY(1–37), which becomes amidated at its COOH-terminal end by peptidyl-glycine-α-amidating monooxygenase that cleaves another amino acid. The resulting fragment, NPY(1–36), is referred to as biologically active NPY or simply NPY (Hodges et al., 2009).

The C-terminal amidation is critically important for the binding of agonists to all Y receptors (Parker and Balasubramaniam, 2008). Residues 1–3 and 33–36 are critically important in the binding of agonists to the Y1 site, and also necessary in the Y4 binding (Keire et al., 2002).

NPY Receptors

The various biological effects of NPY and its homolog are mediated by the activation of at least five receptors, known as the Y1, Y2, Y4, Y5 and y6 subtypes. All of these receptors have been cloned and most are expressed in a range of species. However, the y6 subtype has only been shown to be functionally expressed in mouse and rabbit but not in rat and primates. Existence of a Y3 receptor subtype remains to be established (von Bohlen Und Halbach and Dermietzel, 2006).

NH_2 terminus is essential for NPY to activate Y1 receptor and the complete NPY(1–36) molecule is necessary for NPY binding to Y1 receptor (Michel et al., 1998). Unlike the Y1 receptor, the Y2 receptor accepts as agonists N-terminally truncated NPY and PYY peptides as short as 11 C-terminal residues (Grundemar and Hakanson, 1990). The receptors Y2 and Y5 are equally distantly related to one another as to the Y1/Y4/y6 group. In fact the Y1/Y4/y6 group, the Y2, and the Y5 receptor are more distantly related to one another than any other G-protein-coupled receptors that bind the same endogenous ligand, despite the fact that Y1, Y2, and Y5 all bind two distinct endogenous ligands, namely NPY and PYY (Michel et al., 1998). However, comparing the third intracellular loop, Y1, Y4 and Y5 have more than 40% sequence similarity; Y1 and Y5 genes are transcribed in opposite directions from the same promoter (Herzog et al., 1997). The Y2 receptor seems to be an ancient acquisition from tachykinin group (Larhammar et al., 1992; Parker and Balasubramaniam, 2008.)

Actions of NPY are additionally modified by dipeptidyl peptidase IV. This serine protease cleaves the full length NPY (1–36) to its shorter form, NPY (3–36), which is no longer able to bind to Y1, but retains affinity to all other receptors. Hence, dipeptidyl peptidase IV acts as a natural Y1 receptor antagonist and shifts the activities of NPY to those mediated by Y2 and Y5 receptors (Mentlein, 1999).

Data indicate expression of multiple NPY receptors in different organs. For example, Y1, Y2, Y3, and Y5 were identified in the heart (Protas et al., 2003), Y1, Y2 and Y5 receptors were found in kidneys (Bischoff and·Michel, 1998), and Y1, Y2 and Y4 were observed in gastrointestinal tract (Holzer-Petsche et al., 1991; Fujimiya et al., 2000). There is a complex interaction between Y receptors in the control of different physiological processes. Y1/Y5 receptors are involved in the NPY-induced potentiation of paw edema. Y2 and Y4 receptors play an important role in the regulation of bone homeostasis (Lee et al., 2011) and integrate emotional behavior (Tasan et al., 2009). Activation of both the Y2 and the Y4 receptors, as well as of the cholecystokinin/gastrin and the melanocortin 3 and 4 receptors, should contribute to a composite physiological system controlling satiety (Parker and Balasubramaniam, 2008). The Y4 could be paracrinely mobilized by PP for bone growth (Hosaka et al., 2008), perhaps in conjunction with the very strong attachment of PP (Parker et al., 2001, 2002)

Signal Transduction of NPY Receptors

All known NPY receptors belong to the large superfamily of G-protein-coupled, heptahelical receptors. NPY receptors act via pertussis toxin-sensitive G-proteins, i.e., members of the G_i and G_o family. Activation of NPY receptors in many cells leads to suppression of adenylyl cyclase with subsequent reductions in the production of cAMP and increased phospholipase C activation (Michel et al., 1998). This sequence results in increased intracellular Ca^{2+} and potent, long-lasting vasoconstriction (Ekelund and Erlinge, 1997; Malmstrom, 1997).

Except for adenylyl cyclases, other cascade enzymes also can be involved, for instance, in the system of phospholipase C/inositol triphosphate/ diacylglycerol, which results in activation of protein kinase C (Michel, 1991; Balasubramaniam, 1997). Stimulation of NPY Y2-receptors through pertussis toxin sensitive G-proteins can also inhibit the N-type calcium channels without involvement of protein kinase C (Hodges et al., 2009). Protein kinase C activation by NPY can also inhibit other neuronal calcium channels (e.g. L- and P-type) in neuroendocrine cells (McCullough et al., 1998) and NPY can cause a protein kinase C -dependent increase in the delayed rectifier potassium current. For example, stimulation of NPY receptors in the smooth myocytes induces a twofold increase in the intracellular calcium concentration (Heredia et al., 1992).

Localization of Neuropeptide Y in the Autonomic Nervous System

NPY is widely distributed in the autonomic nervous system. The approximately two thirds of neurons in the mammalian sympathetic ganglia, apart from norepinephrine, contain this peptide. NPY is coreleased with norepinephrine from within peripheral sympathetic nerves that supply blood vessels (Figure). Norepinephrine is contained and released from small dense-core vesicles, whereas NPY is contained within large dense- core vesicles. Also, norepinephrine primarily controls vascular smooth muscle tone during basal conditions, whereas NPY should not; its release only occurs during periods of high stress. NPY produces a slow- acting, potent, and persistent increase in vascular contractile state (Hellstrom, 1987; Lundberg et al., 1994; Ruohonen et al., 2009).

However, NPY acts as a neuromodulator not only in the sympathetic but also in the parasympathetic nervous system. NPY-immunoreactive neurons were also found in three cranial parasympathetic ganglia, the otic, sphenopalatine, and ciliary, in the rat. They were most abundant in the otic ganglion, where they comprised 60-80% of the total ganglion cell population. NPY immunoreactivity was present in approximately 15% and 30% of the neurons in the sphenopalatine and ciliary ganglia, respectively (Leblank et al., 1987). It is interesting that NPY immunoreactivity is detectable in the cell bodies but not on the terminals of ciliary neurons (Leblank and Landis, 1988). In pelvic ganglia, NPY is expressed by all sympathetic and parasympathetic neurons innervating the urinary bladder or rectum of male rats (Keast, 1991). In male rats, most VIP-containing pelvic neurons do not express NPY (Keast, 1991, 1995), whereas in female rats this coexpression is common (Gu et al., 1984; Houdeau et al., 1997)

In the intramural ganglia, NPY was also observed quite often. Thus, for instance, it is found in approximately half of enteric neurons of the submucosal nervous plexus in the mouse, rat, and human, where it is most often co-localized with vasoactive intestinal polypeptide (Peaire et al., 1997; Furness, 2000; Fantaguzzi et al., 2009). In guinea pigs, NPY immunoreactivity was localized in four classes of myenteric neurons: anally projecting interneurons, neurons that projected anally and to the circular muscle, neurons projecting to the longitudinal muscle and in a small population of secretomotor neurons that projected to the mucosa (Uemura et al., 1995). In the intrinsic cardiac ganglia, NPY immunoreactivity was seen in all principal neuronal somata in rats (Richardson et al., 2003). Nevertheless, NPY immunoreactivity was not observed in intracardiac ganglionic neurons in humans (Gordon et al., 1993). It is interesting that human NPY expression could generally be quite less than in rodents, and even in monkeys. Plasma NPY concentrations are much lower in man than in rhesus (Grouzmann et al., 1988; Larsen et al., 1999).

Dense plexuses of NPY immunoreactivity have been identified around coronary vessels and cerebral arteries of cat, guinea pig, rat, and human (McDermott et al., 1993; Gulbenkian et al., 1994; Jansen-Olesen et al., 2004). The peptide is widely distributed in the gastrointestinal tract, in smooth muscles, in muscularis mucosa, and in surrounding blood vessels (Wang et al., 1987). Numerous nerve fibers, immunoreactive for NPY, have been shown to be associated with vascular smooth muscle in the respiratory tract and NPY immunoreactivity has also been found in arteries and arterioles of the lung (Uddman et al., 1984; Kondo et al., 2000). NPY is also located within the male and female reproductive systems (Keator et al., 2010). NPY-immunopositive fibers have been identified in the human fallopian tube, supplying vascular and non-vascular muscle (Heinrich et al., 1987; Markiewicz et al., 2003). In the urinary system, neuropeptide Y-like immunoreactivity has been found throughout the ureter in the vicinity of blood vessels (Allen et al., 1990; Dixon et al., 2000).

Thus, NPY can be localized in populations of neurons with opposite functions, but innervating the same target organ. For instance, this peptide is present in the sympathetic vasoconstrictor and pelvic vasodilator neurons innervating arteries of the uterus in the guinea pig (Morris et al., 1995).

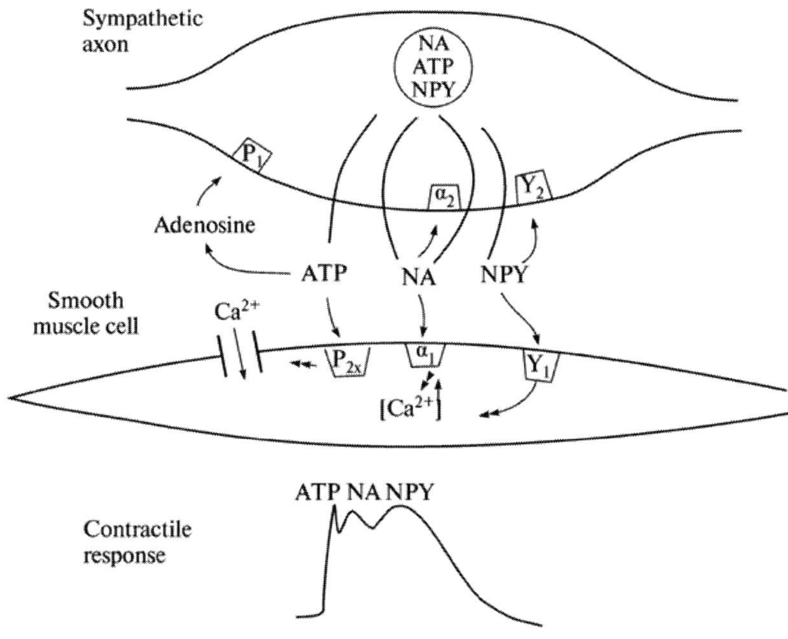

Figure 1. Maslyukov & Nozdrachev Scheme of smooth muscle cell response to ATP, noradrenalin (NA), and neuropeptide Y (NPY) released from sympathetic terminal (Adapted from Morris et al., 1995).

NPY is also found in the sympathetic neurons innervating heart and in parasympathetic neurons of the cardiac intramural ganglia (Richardson et al., 2003).

NPY Expression during the Development in Sympathetic Ganglia

In avians, the initial expression of NPY follows the expression of catecholaminergic properties of neurons (Garcia-Arraras et al., 1992). In bullfrog sympathetic ganglia, NPY expression begins after adrenergic differentiation (Stofer and Horn, 1990). By contrast, in rats, NPY immunoreactivity is first detected in sympathetic ganglia at 12.5 embryonic day (E12.5). When it first appears, NPY is present in almost all tyrosine hydroxylase (TH)-immunopositive cells and the peptide immunofluorescence is faint throughout the cytoplasm (TH is a key enzyme in catecholamine synthesis). During the next 3 days, the proportion of TH-positive cells with peptide-immunoreactivity remains relatively constant while the intensity of peptide immunofluorescence increases. At E16.5, the immunofluorescent intensity of NPY immunoreactivity begins to appear heterogeneous; some cells are more brightly immunoreactive than others. As development proceeds, the proportion of cells with detectable NPY immunoreactivity decreases significantly from approximately 90% in both the superior cervical ganglion and the stellate ganglion at E16.5 to 76% in the superior cervical ganglion and 62% in the stellate ganglion at E18 (Figure). At birth, the proportion of neurons with NPY immunoreactivity in the superior cervical ganglion and in the stellate ganglion is slightly more than 50% (Tyrrell et al., 1992; Tyrrell and Landis, 1994).

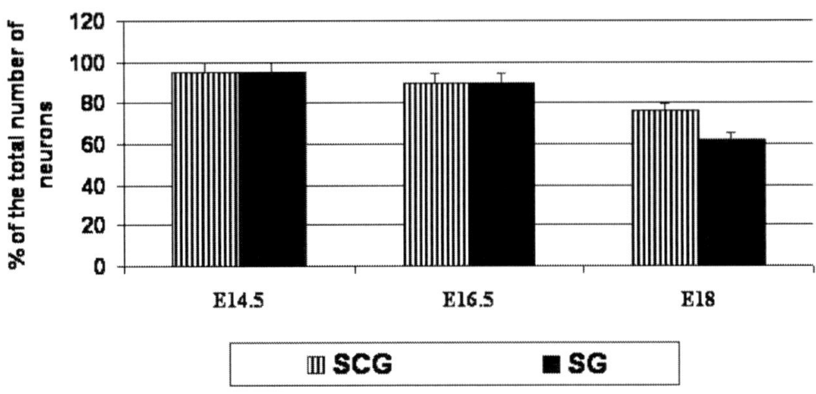

Figure 2. Masliukov & Nozdrachev   Percentage of NPY-immunoreactive neurons in the stellate (SG) and superior cervical ganglion (SCG) of rats during early embryonic development (Masliukov, 2008).

In guinea-pig, NPY was demonstrated in cutaneous sensory and sympathetic neurons during embryonic and fetal development. NPY-immunoreactive axons innervating cutaneous vessels in the ear pinna did not have TH-immunoreactivity in guinea-pig embryos. TH-immunoreactive axons with NPY immunoreactivity appeared later, from midfetal stages (Morris et al., 2001). The proportion of superior cervical ganglion cells with NPY immunoreactivity decreased from 80% at late embryonic stages to the adult level of 50–60% by early fetal stages (Morris et al., 2001). In prevertebral ganglia at Carnegie stages 14–15, NPY is observed in many of the most rostrally located cells, whereas only weak NPY immunoreactivity is present in cells at levels caudal to the developing stomach. By stages 16-17, NPY immunopositive cells are first identified in the caudally located paravertebral chains and in the prevertebral-adrenal regions. NPY immunoreactivity is found in cells with either moderate or intense TH immunoreactivity. From late embryonic period, as many as 80% of TH-immunoreactive neurons have NPY- immunoreactivity. The proportion of TH-immunoreactive neurons with NPY immunoreactivity decreased significantly to 55% by midfetal stages and remained constant through the rest of fetal growth (Morris et al., 2001). In the developing celiac ganglia of guinea-pig, at late embryonic stages, a similar proportion of neurons expresses detectable NPY immunoreactivity in medial and lateral regions of the ganglion. From early fetal stages onward, a proportion of NPY-immunopositive neurons located laterally increases significantly greater than medially located cells (Anderson et al., 2001).

After birth, the proportion of NPY- immunoreactive neurons in the sympathetic ganglia of rats and mice is not constant but increases during the early postnatal period from the moment of birth until the second postnatal month (Figure 3, 4). Nevertheless, there are some differences in the percentage of NPY-positive neurons during the postnatal development of sympathetic ganglia. A larger percentage of NPY-IR neurons was observed in the prevertebral ganglia in comparison with paravertebral ones. The number of NPY-containing neurons is higher in the rat than in the mouse stellate ganglion for all ages in postnatal ontogenesis. Only single NPY- immunopositive and TH- immunonegative neurons are observed in the stellate ganglion of rodents (Masliukov and Timmermans, 2004; Masliukov et al., 2005, 2010).

In the para- and prevertebral ganglia of human fetuses at 24-27 weeks only a small (from 1.5% in celiac ganglia up to 7% in SG) population of NPY-immunoreactive nerve cells are observed. After birth, the number of NPY-positive ganglionic neurons substantially increases and reaches 27% in mesenteric ganglia and 41% in the stellate ganglion. During the development, the proportion of NPY-containing cells continues to increase and reaches 63% in mesenteric ganglia and 73% in the stellate ganglion. In aged humans, the number of NPY- immunoreactive neurons decreases to 37% in mesenteric ganglia and 48% in the stellate ganglion. (Roudenok, 2000).

Sympathetic NPY-immunoreactive nerve fibers showed the earliest expression by 16 days of rat gestation in myocardium (Shoba and Tay, 2000). However, NPY nerve fiber densities were low at birth. Densities increased during first two weeks of postnatal life and then remained relatively stable until adulthood (Nyquist-Battie et al., 1994).

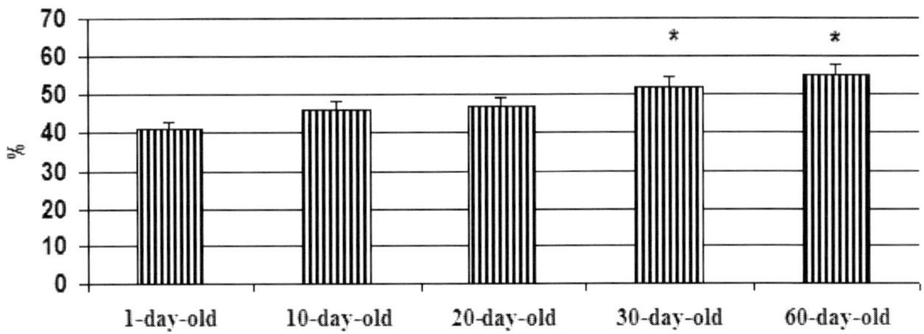

Figure 3. Masliukov & Nozdrachev Percentage of NPY-immunopositive neurons in the stellate ganglion of rats during postnatal development (*P<0.05, comparison was made between newborn and other age groups) (Masliukov, 2005).

Later, developing nerve fibers in the atria of young rats express abundant NPY and this fact would suggest that the peptide may play a trophic role in the development of specialized cardiomyocytes (Nyquist-Battie et al., 1994). In support of a neurotrophic role for NPY, the levels of this peptide are three times higher in the right atrium of immature rats compared to adult rats (Corr et al., 1990). NPY-positive nerve fibers in the mouse liver were first detected in 19-day-old embryos. After birth, the density of positive fibers increased with age to reach maximal level at 1 week but later decreased and reached stable adult levels after 2 weeks postnatal (Ding et al., 1997).

Development of NPY-Immunopositive Parasympathetic Ganglionic Neurons

Unfortunately, we have less data about the development of NPY-immunopositive parasympathetic ganglionic neurons. There are contradictory data about the appearance of NPY in cardiac intramural ganglionic neurons. Horackova et al. (2000) did not observe NPY-immunoreactive neurons in the rat atria during the first two weeks of postnatal development. However, other authors found NPY-immunopositive cardiac neurons even in rat embryos (Shoba and Tay, 2000). NPY immunoreactivity was localized to a large proportion of the intrinsic cardiac ganglia from 16 days of gestation onwards with a progressive increase in the number of neuronal cell bodies per ganglia with age. NPY immunoreactivity appeared in the gut relatively early in ontogeny. It was first detected at day E 12 in the wall of the stomach. By day E 13, NPY immunoreactivity was found in cells in the mesenchyme of the gut at all levels, from the stomach through the colon. By day E 14, NPY immunoreactive cells appeared to begin segregating into distinct layers within the gut wall. By day E 16, NPY immunoreactivity was found in neuritic processes within the enteric plexuses at all levels of the bowel (Branchek and Gershon, 1989).

Expression of Npy Receptors During the Development

NPY and several types of NPY receptors are expressed in the central and the peripheral nervous system at early embryonic stages during nervous system development. Identifying the cells expressing these receptors in both the central and the peripheral nervous system reveals that Y1 and Y2 are expressed in different neurons whereas Y1 and Y5 are often expressed in the same cells (Naveilhan et al., 1998). This would be in agreement with localization on the same chromosome and transcription from the same or similar promoter, located nearby (Herzog et al., 1997).

In whole brain, Y1 and Y2 and Y5 mRNAs were present at very low levels at E12 but were already upregulated a few days later. In the development of sensory and sympathetic ganglia, Y1 upregulation was found to precede Y2. In rats, Y1 mRNA was detected already at E12, a stage characterized by the precursor cell cycle exit and birth of most spinal sensory neurons. The level of Y1 then remained similar in sensory neurons throughout development and into adulthood. Y2 mRNA expression was detected only at E18 and continued to increase until P4 (4th day after birth), at which time adult levels had been reached (Naveilhan et al., 1998). The co-expression of Y1 and Y5 in neurons should reflect a common cis-acting transcriptional control since the genes map to the same locus on chromosome (4q31, and are transcribed in opposite direction from the same promoter (Herzog et al., 1997)

Factors Stimulating NPY Expression

During the embryonic development, NPY phenotype results from a complex combination of regulatory cues; neuronal phenotype is not due solely to environmental cues or lineage restrictions alone, but arises from multiple interactions (Hall and MacPhedran, 1995). Nerve growth factor is required for the maintenance of NPY expression in some neurons (Shadiack et al., 2001). The change in proportion of sympathetic neurons with NPY-IR during embryonic and fetal development may be a consequence of different target tissues. The target tissues produce different levels of nerve growth factor or other trophic factors that affect NPY expression (Rao et al., 1992; Matsumoto et al., 1993). Glucocorticoids and epinephrine are also strong promoters of NPY expression in many cells (Higuchi et al., 1988; Wahlestedt et al., 1990; Han et al., 2012).

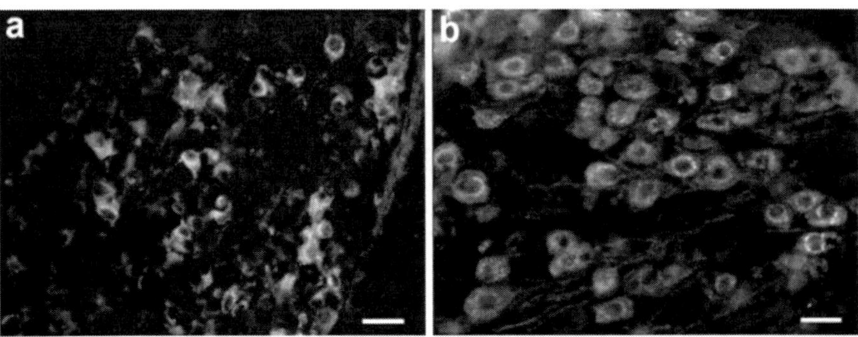

Figure 4. Masliukov & Nozdrachev   Fluorescence micrographs of NPY-IR neurons in the stellate ganglion in newborn (a) and 30-day-old rats (b). Bar, 50 μm (Masliukov, 2008).

The expression of NPY immunoreactivity in celiac ganglionic neurons in guinea-pig fetuses was found to be specifically elevated by culturing the neurons in medium conditioned by dissociated vascular cells (Matsumoto et al., 1993). Rat aortic vascular smooth muscle cells did not affect NPY but tail artery vascular smooth muscle cells increased NPY expression in cultures of dissociated sympathetic neurons (Damon, 2008). Vascular smooth muscle cells expressed leukemia inhibitory factor and neurotrophin-3. Sympathetic neurons innervating blood vessels expressed leukemia inhibitory factor and neurotrophin-3 receptors (Rao et al., 1992; Donovan et al., 1995; Damon, 2008). However, inhibition of leukemia inhibitory factor in aortic neurovascular cocultures did not alter NPY expression (Damon, 2008). Vascular smooth muscle cells-derived neurotrophin-3 maintains TH and NPY expression in the postganglionic sympathetic neurons that innervate them and modulates the function of the postganglionic sympathetic neurons innervating blood vessels (Belliveau et al., 1997; Ramirez-Ordonez et al., 1999; Damon, 2008). Vascular-derived NT-3 is retrogradely transported in postganglionic sympathetic neurons that innervate blood vessels (Zhou et al., 1997)

Vasoactive intestinal peptide (VIP) and pituitary adenylate cyclase activating peptide (PACAP) also stimulate NPY expression. Both VIP and PACAP stimulated expression of the NPY gene through activation of cAMP-dependent protein kinase. PACAP was 1000-fold more potent in eliciting this response compared to VIP (Colbert et al., 1994).

Brain-derived neurotrophic factor (BDNF) increased NPY expression in newborn and adult rat brain. BDNF also increased the number of NPY neurons and promoted the differentiation/maturation of NPY-ergic neurons (Croll et al., 1994; Wirth et al., 1998). However, these increases were smaller than in embryonic neurons. These results suggest that there is a high degree of plasticity of NPY-ergic phenotype in embryonic neurons and that it decreases but persists in postnatal neurons. NPY expression was strongly increased by neurotrophin-4/5 similarly to BDNF and neurotrophin-3 evoked a slight increase (Takei et al., 1996).

Proliferative Action of NPY

NPY may play an important developmental role by promoting growth and/or differentiation of a variety of cells in a receptor-specific manner. Via its Y1R, NPY stimulates cellular proliferation in the mouse central nervous system (Hansel et al., 2001; Decressac et al., 2009), and osteoblasts (Sousa et al., 2009), whereas via the Y1 and Y2R/Y5R it promotes angiogenesis and preadipocyte differentiation (Kuo et al., 2007). NPY has also been shown to stimulate proliferation of human embryonic stem cells mediated by the Y1R and Y5R (Son et al., 2011).

NPY has been shown to be a potent, multifunctional angiogenic factor, which stimulates proliferation, migration, and capillary tube formation in endothelial cells. NPY appears to increase not only capillary angiogenesis but also arteriogenesis or collateral vessel formation (Lee et al., 2003). NPY is angiogenic at concentrations below those required for vasoconstriction (Zukowska-Grojec et al., 1998). NPY is a factor acting upstream from several angiogenic pathways. Activation of these secondary mediators, as well as the direct effect of NPY on endothelial and vascular smooth muscle cells are involved in the multistep process of vascularization resulting in the formation of capillaries and mature arterial vessels (Kitlinska, 2007).

Y1, Y2 and Y5 receptors could cooperate in angiogenesis (Zukowska-Grojec et al., 1998; Parker, Balasubramaniam, 2008; Pons et al., 2008). The early events of vessel formation (adhesion and migration) could be dependent on Y1 and Y2 receptors, which are then markedly upregulated, while the final angiogenic effect may result from increased expression and activation of the Y2 receptors in the later phases. The Y1 receptors stimulate proliferation of endothelial cells and, although to a lesser degree than the Y2 receptors, capillary formation (Zukowska-Grojec et al., 1998). However, vascular smooth muscle cell proliferation appears bimodal, and mediated through 2 receptors, Y1 and Y5. In rat aortic vascular smooth muscle cells, NPY-mediated mitogenesis signals primarily via Y1 receptors activating two Ca^{2+}-dependent growth-promoting pathways , PKC and CaMKII. At the high-affinity peak, these two pathways are amplified by Y5 receptor-mediated, calcium-independent inhibition of the adenylyl cyclase - protein kinase A (PKA) pathway. The three mechanisms converge into the extracellular signal-regulated kinases (ERK1/2) signaling cascade leading to proliferation of vascular smooth muscle cells (Pons et al., 2008).

NPY promotes growth of dorsal root ganglion neurites in vivo and in vitro (White and Mansfield, 1996; White, 1998) and attracts growth cones of embryonic dorsal root ganglion neurons. These effects are mediated via the Y1 receptor (Hökfelt et al., 2008). This attractive effect is stronger than those by nerve growth factor or insulin-like growth factor-1, but less pronounced than that by hepatocyte growth factor (Sanford et al., 2008).

NPY also is associated with cardiac hypertrophy. NPY produces hypertrophy of rat cardiac myocytes by increasing protein synthesis. The effect is age dependent, with a maximum (20% increase of de novo protein synthesis) observed in cells from 16-week-old rats. A release of NPY from sympathetic nerve terminals results in two distinct trophic actions: 1) an increase in L-type Ca^{2+} current density and 2) functional expression of an α-adrenergic signaling cascade whose acute activation is linked to negative chronotropy (Protas et al., 2003).

Conclusion

NPY is a widely distributed neuropeptide in various parts of the autonomic nervous system including the sympathetic and parasympathetic ganglia. Action of NPY is realized through different types of receptors (Y1–Y6) that are located postsynaptically and presynaptically. Apart from effect on vascular tone, heart activity, secretory and motor functions of gastrointestinal tract, NPY plays an important role during the development and produces trophic effects, in particular, stimulates neurogenesis, angiogenesis, and hypertrophy of myocardium. This depends on activity of different NPY receptors. During development, the mainly metabotropic Y1/Y5 pair may overbalance the more structure-linked Y2/Y4 pair, as apparently different from the normal adult equilibrium of the two pairs. The Y2 is involved in bone growth and linking of epithelia, and the Y4 could be paracrinely mobilized by PP for similar tasks (Parker et al., 2001, 2002; Hosaka et al., 2008). Nevertheless, developmental role of NPY still requires considerable clarification.

Acknowledgments

This work was supported by RFBR, a grant of the President of the Russian Federation for young scientists and the Russian Federal Special-Purpose Program "Scientific and scientific-pedagogical personnel of innovative Russia" 2009-2013.

References

Allen JM, Rodrigo J, Kerle DJ, Darcy K, Williams G, Polak JM, and Bloom SR (1990) Neuropeptide Y (NPY)-containing nerves in mammalian ureter. *Urology* 35: 81-86.

Anderson RL, Morris JL, and Gibbins IL (2001) Neurochemical differentiation of functionally distinct populations of autonomic neurons. *J. Comp. Neurol.* 429: 419–435.

Balasubramaniam AA (1997) Neuropeptide Y family of hormones: receptor subtypes and antagonists. *Peptides* 18: 445–457.

Bell D, Allen AR, Kelso EJ, Balasubramaniam A, and McDermott BJ (2002) Induction of hypertrophic responsiveness of cardiomyocytes to neuropeptide Y in response to pressure overload. *J. Pharmacol. Exp. Ther.* 303: 581-591.

Belliveau DJ, Krivko I, Kohn J, LaChance C, Pozniak C, Rusakov D, Kaplan D, and Miller FD (1997) NGF and neurotrophin-3 both activate trk A on sympathetic neurons but differentially regulate survival and neuritogenesis. *J. Cell Biol.* 136: 375–388.

Bischoff A and Michel MC (1998) Renal effects of neuropeptide Y. *Eur. J. Physiol.* 435: 443–453.

Branchek TA and Gershon MD (1989) Time course of expression of neuropeptide Y, calcitonin gene-related peptide, and NADPH diaphorase activity in neurons of the developing murine bowel and the appearance of 5-hydroxytryptamine in mucosal enterochromaffin cells. *J. Comp. Neurol.* 285: 262–273.

Colbert RA, Balbi D, Johnson A, Bailey JA, and Allen JM (1994) Vasoactive intestinal peptide stimulates neuropeptide Y gene expression and causes neurite extension in PC12 cells through independent mechanisms. *J. Neurosci.* 14: 7141-7147.

Corr LA, Aberdeen JA, Milner P, Lincoln L, and Burnstock G (1990) Sympathetic and non-sympathetic neuropeptide Y containing nerves in the rat myocardium and coronary arteris. *Circ. Res.* 66: 1602-1609.

Croll SD, Wiegand SJ, Anderson KD, Lndsay RM, and Nawa H (1994) Regulation of neuropeptides in adult rat forebrain by the neurotrophins BDNF and NGF. *Eur. J. Neurosci.* 6: 1343-1353.

Damon DH (2008) TH and NPY in sympathetic neurovascular cultures: role of LIF and NT-3. *Am. J. Physiol. Cell Physiol.* 294: C306-312.

Decressac M, Prestoz L, Veran J, Cantereau A, Jaber M, and Gaillard A (2009) Neuropeptide Y stimulates proliferation, migration and differentiation of neural precursors from the subventricular zone in adult mice. *Neurobiol. Dis* 34: 441-449.

Ding WG, Kitasato H, and Kimura H (1997) Development of neuropeptide Y innervation in the liver. *Microsc. Res. Tech.* 39: 365-371.

Dixon JS, Jen PY, and Gosling JA (2000) The distribution of vesicular acetylcholine transporter in the human male genitourinary organs and its co-localization with neuropeptide Y and nitric oxide synthase. *Neurourol. Urodyn.* 19: 185-194.

Donovan MJ, Miranda RC, Kraemer R, McCaffrey TA, Tessarollo L, Mahadeo D, Sharif S, Kaplan DR, Tsoulfas P, Parada L, Toran-Allerand CD, Hajjar DP, and Hempstead BL (1995) Neurotrophin and neurotrophin receptors in vascular smooth muscle cells. Regulation of expression in response to injury. *Am. J. Pathol.* 147: 309-324.

Ekelund U and Erlinge D (1997) In vivo receptor characterization of neuropeptide Y-induced effects in consecutive vascular sections of cat skeletal muscle. *Br. J. Pharmacol.* 120: 387–392.

Fujimiya M, Itoh E, Kihara N, Yamamoto I, Fujimura M, and Inui A (2000) Neuropeptide Y induces fasted pattern of duodenal motility via Y(2) receptors in conscious fed rats. *Am. J. Physiol. Gastrointest. Liver Physiol.* 278: G32-38.

Furness JB (2000) Types of neurons in the enteric nervous system. *J. Auton. Nerv. Syst.* 81: 87–96.

Garcia-Arraras JE, Lugo-Chinchilla AM, and Chevere-Colon I (1992) The expression of neuropeptide Y immunoreactivity in the avian sympathoadrenal system conforms with two models of coexpression development for neurons and chromaffin cells. *Development* 115: 617-627.

Gordon L, Polak JM, Moscoso G.J, Smith A, Kuhn DM, and Wharton J (1993) Development of the peptidergic innervation of human heart. *J. Anat.* 183: 131–140.

Grundemar L and Hakanson R (1990) Effects of various neuropeptide Y/peptide YY fragments on electrically- evoked contractions of the rat vas deferens. *Br. J. Pharmacol.* 100: 190–192.

Gu J, Polak JM, Su HC, Blank MA, Morrison JFB, and Bloom SR (1984) Demonstration of paracervical ganglion origin for the vasoactive intestinal peptidecontaining nerves of the rat uterus using retrograde tracing techniques combined with immunocytochemistry and denervation procedures. *Neurosci. Lett.* 51: 377–382.

Gulbenkian S, Saetrum Opgaard O, Barroso CP, Wharton J, Polak JM, and Edvinsson L (1994) The innervation of guinea pig epicardial coronary veins: immunohistochemistry, ultrastructure and vasomotility. *J. Auton. Nerv. Syst.* 47: 201-212.

Hall AK and MacPhedran SE (1995) Multiple mechanisms regulate sympathetic neuronal phenotype. *Development* 121: 2361–2371.

Han R, Kitlinska JB, Munday WR, Gallicano GI, and Zukowska Z (2012) Stress hormone epinephrine enhances adipogenesis in murine embryonic stem cells by up-regulating the neuropeptide y system. *PLoS. One* 7: e36609.

Hansel DE, Eipper BA, and Ronnett GV (2001) Neuropeptide Y functions as a neuroproliferative factor. *Nature* 410: 940-944.

Heinrich D, Reinecke M, Gauwerky JF, and Forssmann WG (1987) Immunohistochemical and biological evidence for a neuromodulator function of neuropeptide Y in the human oviduct. *Arch. Gynecol. Obstet.* 241: 127-132.

Hellstrom PM (1987) Mechanisms involved in colonic vasoconstriction and inhibition of motility induced by neuropeptide Y. *Acta Physiol. Scand.* 129: 549-556.

Heredia MP, Fernandez-Velasco M, Benito G, and Delgado C (2002) Neuropeptide Y increases 4-aminopyridine-sensitive transient outward potassium current in rat ventricular myocytes. *Br. J. Pharmacol.* 135: 1701–1706.

Herzog H, Darby K, Ball H, Hort Y, Beck-Sickinger A, and Shine J (1997) Overlapping gene structure of the human neuropeptide Y receptor subtypes Y1 and Y5 suggests coordinate transcriptional regulation. *Genomics* 41: 315-319.

Higuchi H, Yang HY, and Sabol SL (1988) Rat neuropeptide Y precursor gene expression mRNA structure, tissue distribution, and regulation by glucocorticoids, cyclic AMP, and phorbol ester. *J. Biol. Chem.* 263: 6288–6295.

Hodges GJ, Jackson DN, Mattar L, Johnson JM, and Shoemaker JK (2009) Neuropeptide Y and neurovascular control in skeletal muscle and skin. *Am J. Physiol. Regul. Integr. Comp. Physiol.* 297: 546-555.

Hökfelt T, Stanic D, Sanford SD, Gatlin JC, Nilsson I, Paratcha G, Ledda F, Fetissov S, Lindfors C, Herzog H, Johansen JE, Ubink R, and Pfenninger KH (2008) NPY and its involvement in axon guidance, neurogenesis, and feeding. *Nutrition* 24: 860-868.

Holzer-Petsche U, Petritsch W, Hinterleitner T, Eherer A, Sperk G, and Krejs GJ (1991) Effect of neuropeptide Y on jejunal water and ion transport in humans. *Gastroenterology* 101: 325–330.

Horackova M, Slavikova J, and Byczko Z (2000) Postnatal development of the rat intrinsic cardiac nervous system: a confocal laser scanning microscopy study in whole-mount atria. *Tissue Cell* 32: 377–388.

Hosaka H, Nagata A, Yoshida T, Shibata T, Nagao T, Tanaka T, Saito Y, and Tatsuno I (2008) Pancreatic polypeptide is secreted from and controls differentiation through its specific receptors in osteoblastic MC3T3-E1 cells. *Peptides* 29:1390-1395.

Houdeau E, Prud'homme MJ, Rousseau A, and Rousseau JP (1995) Distribution of noradrenergic neurons in the female rat pelvic plexus and involvement in the genital tract innervation. *J. Auton. Nerv. Syst.* 54: 113–125.

Jansen-Olesen I, Gulbenkian S, Engel U, Cunha SM, and Edvinsson L (2004) Peptidergic and non-peptidergic innervation and vasomotor responses of human lenticulostriate and posterior cerebral arteries. *Peptides* 25: 2105-2114.

Keast JR (1991) Patterns of co-existence of peptides and differences of nerve fibre types associated with noradrenergic and non-noradrenergic (putative cholinergic) neurons in the major pelvic ganglion of the male rat. *Cell Tissue Res* 266: 405–415.

Keast JR (1995) Visualization and immunohistochemical characterization of sympathetic and parasympathetic neurons in the male rat major pelvic ganglion. *Neuroscience* 66: 655–662.

Keator CS, Custer EE, Hoagland TA, Schreiber DT, Mah K, Lawson AM, Slayden OD, and McCracken JA (2010) Evidence for a potential role of neuropeptide Y in ovine corpus luteum function. *Domest. Anim. Endocrinol.* 38: 103-114.

Keire DA, Bowers CW, Solomon TE, and Reeve Jr JR (2002) Structure and receptor binding of PYY analogs. *Peptides* 23: 305–321.

Kitlinska J (2007) Neuropeptide Y (NPY) in neuroblastoma: effect on growth and vascularization. *Peptides* 28:405-412.

Kondo T, Inokuchi T, Ohta K, Annoh H, and Chang J (2000) Distribution, chemical coding and origin of nitric oxide synthase-containing nerve fibres in the guinea pig nasal mucosa. *J. Auton. Nerv. Syst.* 80: 71-79.

Kuo LE, Kitlinska JB, Tilan JU, Li L, Baker SB, Johnson MD, Lee EW, Burnett MS, Fricke ST, Kvetnansky R, Herzog H, and Zukowska Z (2007) Neuropeptide Y acts directly in the periphery on fat tissue and mediates stress-induced obesity and metabolic syndrome. *Nat. Med.* 13: 803–811.

Larhammar D, Blomqvist AG, Yee F, Jazin E, Yoo H, and Wahlested C (1992) Cloning and functional expression of a human neuropeptide Y/peptide YY receptor of the Y1 type. *J. Biol. Chem.* 267: 10935-10938.

Leblanc GG and Landis SC (1988) Target specificity of neuropeptide Y-immunoreactive cranial parasympathetic neurons. *J. Neurosci.* 8: 146-155.

Leblanc GG, Trimmer BA, and Landis SC (1987) Neuropeptide Y-like immunoreactivity in rat cranial parasympathetic neurons: Coexistence with vasoactive intestinal peptide and choline acetyltransferase. *Proc. Nati. Acad. Sci. USA* 84: 3511-3515.

Lee EW, Michalkiewicz M, Kitlinska J, Kalezic I, Switalska H, Yoo P, Sangkharat A, Ji H, Li L, Michalkiewicz T, Ljubisavljevic M, Johansson H, Grant DS, and Zukowska Z (2003) Neuropeptide Y induces ischemic angiogenesis and restores function of ischemic skeletal muscles. *J. Clin. Invest.* 111: 1853–1862.

Lee NJ, Allison S, Enriquez RF, Sainsbury A, Herzog H, and Baldock PA (2010) Y2 and Y4 receptor signalling attenuates the skeletal response of central NPY. *J. Mol. Neurosci.* 43: 123-131.

Lundberg JM, Franco-Cereceda A, Lou YP, Modin A, and Pernow J (1994) Differential release of classical transmitters and peptides. *Adv. Second. Messenger. Phosphoprotein Res.* 29: 223–234.

Malmstrom RE (1997) Neuropeptide Y Y1 receptor mechanisms in sympathetic vascular control. *Acta Physiol. Scand Suppl.* 636: 1–55.

Markiewicz W, Jaroszewski JJ, Bossowska A, and Majewski M (2003) NPY: its occurrence and relevance in the female reproductive system. *Folia Histochem. Cytobiol.* 41: 183-192.

Masliukov PM (2008) Development of neurotransmitter content in sympathetic ganglia, in *Sympathetic nervous system research development* (Kaneko M ed) pp 127-146, Nova Science Publishers, New York.

Masliukov PM and Timmermans JP (2004) Immunocytochemical properties of stellate ganglion neurons during early postnatal development. *Histochem. Cell Biol.* 122: 201-209.

Masliukov PM, Shilkin VV, and Timmermans JP (2005) Immunocytochemical characteristic of neurons of the mouse truncus sympaticus stellate ganglion in postnatal ontogenesis. *Morfologiia* 128: 41-44.

Masliukov PM, Korzina MB, Emanuilov AI, and Shilkin VV (2010) Neurotransmitter composition of neurons in the cranial cervical and celiac sympathetic ganglia in postnatal ontogenesis. *Neurosci. Behav. Physiol.* 40: 143-147.

Matsumoto SG, Gruener RP, and Kreulen DL (1993) Neurotransmitter properties of guinea-pig sympathetic neurons grown in dissociated cell culture-II. Fetal and embryonic neurons: regulation of neuropeptide Y expression. *Neuroscience* 57: 1147–1157.

McCullough LA, Egan TM, and Westfall TC (1998) Neuropeptide Y inhibition of calcium channels in PC-12 pheochromocytoma cells. *Am. J. Physiol.* 274: C1290–1297.

McDermott BJ, Millar BC, and Piper HM (1993) Cardiovascular effects of neuropeptide Y: receptor interactions and cellular mechanisms. *Cardiovasc. Res.* 27: 893-905.

Mentlein R (1999) Dipeptidyl-peptidase IV (CD26)—role in the inactivation of regulatory peptides. *Regul. Pept.* 85: 9–24.

Michel M (1991) Receptors for neuropeptide Y multiple subtypes and multiple second messengers. *Trends Pharmacol. Sci.* 12: 389-394.

Michel MC, Beck-Sickinger A, Cox H, Doods HN, Herzog H, Larhammar D, Quirion R, Schwartz T, and Westfall T (1998) XVI. International Union of Pharmacology recommendations for the nomenclature of neuropeptide Y, peptide YY, and pancreatic polypeptide receptors. *Pharmacol. Rev.* 50: 143–150.

Minth CD, Bloom SR, Polak JM, and Dixon JE (1984) Cloning, characterization, and DNA sequence of a human cDNA encoding neuropeptide tyrosine. *Proc. Natl. Acad. Sci. USA* 81: 4577–4581.

Mongardi Fantaguzzi C, Thacker M, Chiocchetti R, and Furness JB (2009) Identification of neuron types in the submucosal ganglia of the mouse ileum. *Cell Tissue Res.* 336: 179-189.

Morris JL, Anderson RL, and Gibbins IL (2001) Neuropeptide Y immunoreactivity in cutaneous sympathetic and sensory neurons during development of the guinea pig. *J. Comp. Neurol.* 437: 321–334.

Morris JL, Gibbins IL, Kadowitz PJ, Herzog H, Kreulen DL, Toda N, and Claing A (1995) Roles of peptides and other substances in cotransmission from vascular autonomic and sensory neurons. *Can. J. Physiol. Pharmacol.* 73: 521-532.

Naveilhan P, Neveu I, Arenas E, and Ernfors P (1998) Complementary and overlapping expression of Y1, Y2 and Y5 receptors in the developing and adult mouse nervous system. *Neuroscience* 87: 289-302.

Nyquist-Battie C, Cochran PK, Sands SA, and Chronwall BM (1994) Development of neuropeptide Y and tyrosine hydroxylase immuno-reactive innervation in postnatal rat heart. *Peptides* 15: 1461-1469.

O'Hare MMT, Tenmoku S, Aakerlund L, Hilsted L, Johnsen A, and Schwartz TW (1988) Neuropeptide Y in guinea-pig, rabbit, rat and man. Identical amino acid sequence and oxidation of methionine-17. *Regul. Peptides* 20: 293-305.

Parker MS, Berglund MM, Lundell I, and Parker SL (2001) Blockade of pancreatic polypeptide-sensitive neuropeptide Y (NPY) receptors by agonist peptides is prevented by modulators of sodium transport. Implications for receptor signaling and regulation. *Peptides* 22: 887-898.

Parker MS, Lundell I, and Parker SL (2002) Pancreatic polypeptide receptors: affinity, sodium sensitivity and stability of agonist binding. *Peptides* 23: 291-303.

Parker SL and Balasubramaniam A (2008) Neuropeptide Y Y2 receptor in health and disease. *Br. J. Pharmacol.* 153: 420-431.

Peaire AE, Krantis A, and Staines W (1997) Distribution of the NPY receptor subtype Y1 within human colon: evidence for NPY targeting a subpopulation of nitrergic neurons. *J. Auton. Nerv. Syst.* 67: 168–175.

Pons J, Kitlinska J, Jacques D, Perreault C, Nader M, Everhart L, Zhang Y, and Zukowska Z (2008) Interactions of multiple signaling pathways in neuropeptide Y-mediated bimodal vascular smooth muscle cell growth. *Can. J. Physiol. Pharmacol.* 86: 438-448.

Protas L, Qu J, and Robinson RB (2003) Neuropeptide y: neurotransmitter or trophic factor in the heart? *News Physiol. Sci.* 18: 181-185.

Ramirez-Ordonez R, Barreto-Estrada JL, and Garcia-Arraras JE (1999) Growth factors effects on the expression of morphological and biochemical properties of avian embryonic sympathetic cells. Emphasis on NGF. *Dev. Brain Res.* 114: 27–36.

Rao MS, Tyrrell S, Landis SC, and Patterson PH (1992) Effects of ciliary neurotrophic factor (CNTF) and depolarization on neuropeptide expression in cultured sympathetic neurons. *Dev. Biol.* 150: 281–293.

Richardson RJ, Grkovic I, and Anderson CR (2003) Immunohistochemical analysis of intracardiac ganglia of the rat heart. *Cell Tissue Res.* 314: 337–350.

Roudenok V (2000) Changes in the expression of neuropeptide Y (NPY) during maturation of human sympathetic ganglionic neurons: correlations with tyrosine hydroxylase immunoreactivity. *Ann. Anat.* 182: 515-519.

Ruohonen ST, Savontaus E, Rinne P, Rosmaninho-Salgado J, Cavadas C, Ruskoaho H, Koulu M, and Pesonen U (2009) Stress-induced hypertension and increased sympathetic activity in mice overexpressing neuropeptide Y in noradrenergic neurons. *Neuroendocrinology* 89: 351-360.

Sanford SD, Gatlin JC, Hökfelt T, and Pfenninger KH (2008) Growth cone responses to growth and chemotropic factors. *Eur. J. Neurosci.* 28: 268-278.

Shadiack, AM, Sun, YI, and Zigmond RE (2001) Nerve growth factor antiserum induces axotomy-like changes in neuropeptide expression in intact sympathetic and sensory neurons. *J. Neurosci.* 21: 363–371.

Shoba T and Tay SSW (2000) Nitrergic and peptidergic innervation in the developing rat heart. *Anat. Embryol.* 201: 491–500.

Son MY, Kim MJ, Yu K, Koo DB, and Cho YS (2011) Involvement of neuropeptide Y and its Y1 and Y5 receptors in maintaining self-renewal and proliferation of human embryonic stem cells. *J. Cell Mol. Med.* 15: 152–165.

Sousa DM, Herzog H, and Lamghari M (2009) NPY signalling pathway in bone homeostasis: Y1 receptor as a potential drug target. *Curr. Drug Targets* 10: 9–19.

Stofer WD and Horn JP (1990) Expression of neuropeptide-Y-like immunoreactivity begins after adrenergic differentiation and ganglionic synaptogenesis in developing bullfrog sympathetic neurons. *J. Neurosci.* 10: 3305-3312.

Takei N, Sasaoka K, Higuchi H, Endo Y, and Hatanaka H (1996) BDNF increases the expression of neuropeptide Y mRNA and promotes differentiation/maturation of neuropeptide Y-positive cultured cortical neurons from embryonic and postnatal rats. *Brain Res. Mol. Brain Res.* 37: 283-289.

Tasan RO, Lin S, Hetzenauer A, Singewald N, Herzog H, and Sperk G (2009) Increased novelty-induced motor activity and reduced depression-like behavior in neuropeptide Y (NPY)-Y4 receptor knockout mice. *Neuroscience* 158: 1717-1730.

Tatemoto K (1982) Neuropeptide Y. Complete amino acid sequence of the brain peptide. *Proc. Natl. Acad. Sci. USA* 79: 5485–5489.

Tatemoto K, Carlquist M, and Mutt V (1982) Neuropeptide Y—a novel brain peptide with structural similarities for peptide YY and pancreatic polypeptide. *Nature* 296: 659–660.

Tyrrell S and Landis SC (1994) The appearance of NPY and VIP in sympathetic neuroblasts and subsequent alterations of their expression. *J. Neurosci.* 14: 4529-4547.

Tyrrell S, Siegel RE, and Landis SC (1992) Tyrosine hydroxylase and neuropeptide Y are increased in ciliary ganglia of sympathectomized rats. *Neuroscience* 47: 985-998.

Uddman R, Sundler F, and Emson P (1984) Occurrence and distribution of neuropeptide-Y-immunoreactive nerves in the respiratory tract and middle ear. *Cell Tissue Res.* 237: 321-327.

Uemura S, Pompolo S, and Furness JB (1995) Colocalization of neuropeptide Y with other neurochemical markers in the guinea-pig small intestine. *Arch. Histol. Cytol.* 58: 523-536.

von Bohlen und Halbach O and Dermietzel R (2006) Neurotransmitters and neuromodulators. *Handbook of receptors and biological effects.* Wiley-VCH Verlag, Weinheim.

Wahlestedt C, Hakanson R, Vaz CA, and Zukowska-Grojec Z (1990) Norepinephrine and neuropeptide Y: vasoconstrictor cooperation in vivo and in vitro. *Am. J. Physiol.* 258: R736–742.

Wang YN, McDonald JK, and Wyatt RJ (1987) Immunocytochemical localization of neuropeptide Y-like immunoreactivity in adrenergic and non-adrenergic neurons of the rat gastrointestinal tract. *Peptides* 8: 145-151.

White DM (1998) Contribution of neurotrophin-3 to the neuropeptide Y–induced increase in neurite outgrowth of rat dorsal root ganglion cells. *Neuroscience* 86: 257– 263.

White DM and Mansfield K (1996) Vasoactive intestinal polypeptide and neuropeptide Y act indirectly to increase neurite outgrowth of dissociated dorsal root ganglion cells. *Neuroscience* 73: 881–887.

Wirth MJ, Obst K, and Wahle P (1998) NT-4/5 and LIF, but not NT-3 and BDNF, promote NPY mRNA expression in cortical neurons in the absence of spontaneous bioelectrical activity. *Eur. J. Neurosci.* 10: 1457–1464.

Zhou XF, Chie ET, Deng YS, and Rush RA (1997) Rat mature sympathetic neurones derive neurotrophin 3 from peripheral effector tissues. *Eur. J. Neurosci.* 9: 2753–2764.

Zukowska-Grojec Z, Karwatowska-Prokopczuk E, Rose W, Rone J, Movafagh S, Ji H, Yeh Y, Chen WT, Kleinman HK, Grouzmann E, and Grant DS (1998) Neuropeptide Y: a novel angiogenic factor from the sympathetic nerves and endothelium. *Circ. Res.* 83: 187-195.

In: Neuropeptide Y
Editors: Steven L. Parker

ISBN: 978-1-62618-421-3
© 2013 Nova Science Publishers, Inc.

Chapter V

Neuropeptide Y in Feeding: Specific and Protean Actions

*Michael S. Parker[1], Renu Sah[2],
Ambikaipakan Balasubramaniam[3], Floyd R. Sallee[2],
William R. Crowley[4] and Steven L. Parker[5]**

[1]Department of Microbiology and Molecular Cell Sciences,
University of Memphis, Memphis, TN, US
[2]Department of Psychiatry, University of Cincinnati
School of Medicine, Cincinnati, OH, US
[3]Division of Gastrointestinal Hormones, Department of Surgery,
University of Cincinnati School of Medicine, Cincinnati, OH, US
[4]Department of Pharmacology, School of Pharmacy,
University of Utah, Salt Lake City, UT, US
[5]Department of Pharmacology,
University of Tennessee College of Medicine, Memphis, TN, US

Abstract

Neuropeptide Y (NPY) participates in the regulation of aminergic transmission, glucose homeostasis, secretion of hormonal and other peptides, and activity of channels and transporters. Connected to these

---

* The corresponding author, email stevenleonardparker@msn.com.

processes, NPY affects conditions related to feeding, hypertension, diabetes and angiogenesis. These multiple activities are supported by two cooperating NPY receptors, the Y1 and Y5 subtypes, and by the Y2 receptor (possibly in association with the Y4 receptor) somewhat opposing the Y1 and Y5 (especially in the mammal). This counteraction is aided by the proteolytic clipping of Y1/Y5/ Y2-active 36-residue Y peptides to Y2-specific 3-36 peptides. This chapter re-examines the properties of brain NPY receptors and presents new evidence for receptor subtype- and brain area-related differences in transduction via Y1 and Y2 receptors.

The broad, protean interactions of the 10-16 EDAPAED section of NPY could be important in the observed blockade of the Y1 receptor by NPY and interactions with transducers and effectors (including the voltage sensors of ion channels).

The homobasic 19-36 sector of NPY could interact with extracellular domains of channels. The G-protein-coupling intracellular domains of Y receptors could also couple with effectors, including transporters and channels.

With NPY as ligand (but much less with the related peptide YY), the seesaw feeding regulation via Y1/Y5 vs. Y2/(Y4) receptors could significantly depend on these additional interactions.

Keywords: Y1 receptor; Y2 receptor; receptor masking; receptor blockade; homoacidic tract; homobasic tract

Abbreviations

aa	amino acid(s) or amino acid residues;
csf	the cerebrospinal fluid; ec, extracellular domain;
GβS	guanosine-5'-(β-thio)diphosphate;
GγS	guanosine-5'-(γ-thio)triphosphate;
GPCR	G-protein coupling receptor;
h	human;
ic	intracellular domain;
icv	injection into the third cerebral ventricle;
NPY	neuropeptide Y;
p	porcine;
Pdb	Protein Data Bank;
PYY	peptide YY;
PP	pancreatic polypeptide;

PrlRP   the prolactin-releasing peptide;
PTX   pertussis toxin;
r   rat;
Snp   single nucleotide polymorphism;
tm   transmembrane domain;
Y peptide   a NPY, PYY or PP-like peptide;
Y receptor   a NPY, PYY or PP receptor;
Y1R,Y2R, Y4R, Y5R, the respective Y receptors

Introduction

Three major regulators of glucose homeostasis and feeding, insulin, leptin and NPY, all act on multiple targets. The targets include primary specific receptors (which are not very numerous), many associate receptors as secondary sites (for insulin e.g. those for the insulin-like growth factors), and medium to low-affinity tertiary sites (for insulin, a number of receptor tyrosine kinases).

This protean hierarchy also depends on transducers and effectors, making it difficult to directly identify the partners of the three hormones, and in turn complicates interpretation of the feeding effects. Due to these problems, effects of insulin and leptin are rarely evaluated in connection to receptor binding.

However, the high binding affinity of the primary Y1 and Y2 receptors for NPY (Aakerlund et al., 1990; Dumont et al., 1993; Larsen et al., 1993; Lundell et al., 1995; Parker et al., 1996a) and the physiological differentiation of NPY-related agonist peptides (Daniels et al., 1995; Grandt et al., 1992) permit reliable estimates and correlation with effects upon feeding.

A word about multi-faceted interactions of peptides and proteins could be useful to avoid a semantic confusion. The protean interactions of receptors usually are taken to refer to different efficacy with similar effectors (Neubig, 2007).

However, the protean interactions of peptidic agonists in a broader sense could involve any specific association with a partner or class of partners, and for peptides with large interaction spectra, such as NPY, this is a logical direction to explore.

1. Properties and Interactions of Y Peptides and Receptors

1.1. The Phylogeny of NPY and NPY Receptors

Neuropeptide Y (NPY) is an ancient 36-peptide of the vertebrate (Larhammar, 1996), made mainly in neural cells and abundant especially in the forebrain (Adrian et al., 1983; Allen et al., 1983). NPY has a C-terminal arginine / amidated aromatic dipeptide (Arg-Tyr.NH$_2$) similar with the RF amide group of peptides. RFamides are however also found in prokaryotes, indicating a broad functional versatility of this C-terminal motif.

This may, but does not have to, have phylogenetic links. The presumed invertebrate relatives of NPY, such as insect neuropeptides F (Table 1), are RFamides; phenylalanine and tryptophan amides terminate chains of some poikilotherm NPY-related 36-peptides (e.g. u-p access P06306, P31229 and P15427).

RFamides (often, unlike NPY, FMRFamides) are found in most multi-cellular eukaryotes, from Hydra (Koizumi et al., 1989) to man, with activities ranging from support of reproduction (Kriegsfeld, 2006) to cardioexcitation (Moulis, 2006), to negative (Cline et al., 2009) or positive (Christie et al., 2011) regulation of feeding, and even to neurotoxicity (Maillo et al., 2002). The C-termini of many RFamides and of all NPY analogues have arginine at C-terminal positions -1 and -3, and the latter represent long ionic zippers (Table 1).

NPY is documented in a host of vertebrates and could also be present in primary chordates (Castro et al., 2003). Similar peptides are found in mollusks and insects (Table 1).

Invertebrate NPF peptides similar in size to NPY could represent a parallel and independent evolution, and this could also apply to vertebrate NPFF peptides. NPY presents a structure with sharp segregation of ionizable residues into an acidic sidechain-rich C-terminal half, and a basic residue-rich N-terminus ending in a RFamide dipeptide (Table 1); for a recent review of structure and interactivity of NPY see (Parker et al., 2011).

The NPY genes have four exons in the vertebrate chromosomal DNAs characterized thus far. Exon 2 contains code for the signal peptide and most of NPY proper (see GenBank records AC_000068 (human), AC_000161 (bovine), NC_006089 (chick) and NC_007130 (fish)). The exons are located considerably apart, which increases the probability of polymorphisms.

Table 1. NPY and similar peptides in eukaryotes

Peptide and species	1 10 20 30	Access *
NPY		
Human/rodent	YPSKPDNPGEDAPAEDMARYYSALRHYINLITRQRY	P01303
Pig/dog	YPSKPDNPGEDAPAEDLARYYSALRHYINLITRQRY	P01304
Sheep	YPSKPDNPGDDAPAEDLARYYSALRHYINLITRQRY	P14765
Chick	YPSKPDSPGEDAPAEDMARYYSALRHYINLITRQRY	P28673
Common frog	YPSKPDNPGEDAPAEDMAKYYSALRHYINLITRQRY	P69102
Atlantic cod	YPIKPENPGEDAPADELAKYYSALRHYINLITRQRY	P80167
PYY		
Human	YPIKPEAPGEDASPEELNRYYASLRHYLNLVTRQRY	P10082
Pig/rat	YPAKPEAPGEDASPEELSRYYASLRHYLNLVTRQRY	P10631
Rabbit	YPSKPEAPGEDASPEELNRYYASLRHYLNLVTRQRY	Q9TR93
Trout	YPPKPENPGEDAPPEELAKYYTALRHYINLITRQRY	P69093
PP		
Human	APLEPVYPGDNATPEQMAQYAADLRRYINMLTRPRY	P01298
Rabbit	APPEPVYPGDDATPEQMAEYVADLRRYINMLTRPRY	P41336
Rat	APLEPMYPGDYATHEQRAQYETQLRRYINTLTRPRY	P06303
Turkey	GPSQPTYPGDDAPVEDLIRFYDNLQQYLNVVTRHRY	P68249
Common frog	APSEPHHPGDQATQDQLAQYYSDLYQYITFVTRPRF	P31229
NPF		
Turbellarian	KVVHLRPRSSFSSEDEYQIYLRNVSKYIQLYGRPRF	P41334
Aedes mosquito	SFTDARPQDDPTSVAEAIRLLQELETKHAQHARPRF	Q8MP00
Fruit fly	NDVNTMADAYKFLQDLDTYYGDRARVRF	Q9VET0
Anopheles	RPQDSDAASVAAAIRYLQELETKHAQHARPRF	Q7Q7R8
Shrimp	KPDPSQLANMAEALKYLQELDKYYSQVSRPRF	F6KM62
NPFF		
Human-37	SLNFEELKDWGPKNVIKMSTPAVNKMPHSFANLPLRF	Q9HCQ7
PrlRP		
Human	SRTHRHSMEIRTPDINPAWYASRGIRPVGRF	P81277

*UniProtein database identifier.

The sequence of human NPY is identical to that of rabbit, rodent (rat, mouse and
guinea-pig cloned) and alligator NPY, while that of pig PYY is the same as for rat
PYY. The non-conservative differences are shown in boldface, while the

conservative differences are represented in italics. The C-terminal residue is amidated in all peptides.

However, sequences of NPY peptides are very highly conserved (Table 1); the homobasic 19-36 section shows a single conservative difference from fish to primate, and none in the mammal (Table 1). Numerous single nucleotide polymorphisms (snps) are found in other sections of NPY genes, especially those coding for the signal peptide.

The signal peptide snp T>C results in mutation L7>P7 which is involved in type 2 diabetes (Nordman et al., 2005), possibly related to increased secretion of NPY (Mitchell et al., 2008). The snp rs17376826 (C>T) in the human Y2R gene could be related to density of the Na^+/K^+ ATPase pump in red blood cells (Hasstedt et al.).

The vertebrate NPY receptors seem to have had uneven phylogenetic history of expression and retention, especially in the fish (Salaneck et al., 2008).

Some NPY receptors may have been lost and re-acquired repeatedly (Starback et al., 2000) through large-scale chromosome modification events. This underlines the apparent functional redundancy of Y receptors, especially in regard to the regulation of feeding. This regulation could be attenuated in poikilotherm vertebrates (Salaneck et al., 2008), but is prominent in birds and mammals, and especially in rodents (Kanatani et al., 2000; Lecklin et al., 2003; Stanley and Leibowitz, 1985).

1.2. Evolution of Selective Agonism in NPY Receptors

Functions of NPY receptor subtypes have differentiated considerably in land animals compared to fishes. The NPY receptor complement varies considerably in fishes (Salaneck et al., 2008), but species that maintain the Y1 gene also show the feeding response to NPY. NPY stimulates feeding in amphibians (Crespi et al., 2004), which have both Y1 and Y5 receptors (Larhammar and Salaneck, 2004). A considerable functional differentiation apparently developed in mammals relative to birds (Berglund et al., 2002; Salaneck et al., 2000).

This seems to especially involve the Y2 receptor, which in mammals in many respects acts as an antagonist of the Y1 receptor. In birds the Y2 receptor is less antagonistic to the Y1 ((Berglund et al., 2002); also see the chapter by Bungo and colleagues in this book). The Y4 agonism also differentiated sharply in the mammal (with very low affinity for NPY (Lundell et al., 1995; Parker et al., 2002a)) compared to birds (with a high affinity for NPY (Lundell et al., 2002)).

1.3. The Binding of NPY and Related Peptides to NPY Receptors

The C-terminal hexad of Y peptides (Table 1) is essential for the specific binding to all Y receptors (Bader et al., 2002; Beck-Sickinger et al., 1994; Gehlert et al., 1996; Kirby et al., 1993). The Y1 receptor also requires the N-terminal dipeptide of NPY for a two-prong discontinuous binding (Daniels et al., 1995), which is a strict, but not an absolute, requirement (O'Shea et al., 1997). These residues are not critical for the Y2 binding, and the affinity of (1-36) and (3-36) NPY and PYY peptides for the Y2 receptor is similarly high (see (Parker and Balasubramaniam, 2008)). The C-terminal hexads of Y peptides are quite similar (Table 1). However, the LTRPRY hexad of pancreatic polypeptides (Table 1) confers selectivity for the mammalian Y1 receptor group (Y1, Y4 and Y5 receptors; see also section 2.2 and Figure 4), and there is Y1 binding of (Pro34)NPY(3-36) (with ITRPRY hexad) at a reduced affinity (O'Shea et al., 1997). The 10-16 section does not participate in the specific binding (Kirby et al., 1995), but strongly influences other attachments. The EDAPAED sector of NPY should support, through the APA and the double ED motifs, a more extensive cross-reactivity than the EDASPEE motif of PYY (Parker et al., 2011). This sector could also support the much higher resistance of NPY binding to chaotropes compared to PYY (Parker et al., 2011; Parker and Parker, 2000; Sah et al., 2005), and could also be important for differences in agonist activity between NPY and PYY (Parker et al., 2007a; Sah et al., 2005).

1.4. Polymorphisms and Variants of Y Receptors and Feeding

The human Y1R mRNA has two coding exons (nucleotides 6242-6940 and 7038-7493), separated by a short intron between transmembrane (tm) tm5 and tm6 sectors (Herzog et al., 1993), and the Y2, Y4 and Y5 mRNAs have single coding exons (nucleotides 5312-6457, 3251-4378 and 6336-7673, respectively). The intracellular (ic) ic3 and ic4 domains of Y1, Y2 and Y4 receptors are moderate in size. The Y5 mRNA has a large ic3 domain (>130 aa), with considerable variations across species (see (Holmberg et al., 2002)).

The Y1 and Y2 receptors are conserved at more than 90% in mammals, as different from the Y5 and especially the Y4 receptors (conserved at less than 70%; (Wraith et al., 2000)). This would predict a large inter-species similarity in the regulation of energy output and feeding for the mammalian Y1 subtype, and also for the Y1-opposing activity of the mammalian Y2 subtype. Reflecting this high conservation, there is thus far no evidence for major isoforms of Y1 and Y2 receptors in the mammal.

The Y1 variants could be abundant in fishes (Lundell et al., 1997), and there is loss of Y1 and Y5-like receptors in several teleost species (see e.g. (Salaneck et al., 2008)), but the positive regulation of feeding by NPY and Y1-type receptors is a rule rather than an exception across the fish genera (Yokobori et al., 2012). Among mammals, an embryonic isoform of the Y1 receptor which does not transduce to Gi/o and Gq was detected in the mouse (Nakamura et al., 1995). A hypothalamic Y1-like (Gi-coupling, pertussis toxin (PTX)-sensitive) receptor was cloned(Weinberg et al., 1996), but thus far is not characterized in terms of occurrence across species, abundance relative to the main (cortical-type) Y1 isoform, and interaction with effectors.

The Y2-like receptors in the fish could be present as major isoforms and closely related subtypes (Larhammar and Salaneck, 2004), but do not seem to inhibit feeding (Aldegunde and Mancebo, 2006). Bird Y2 receptors also have not been reported to serve in reduction of feeding (see the chapter by Bungo and colleagues in this book) , in accord with sequence differences that confer sensitivity to the mammalian Y1-selective ligands (Berglund et al., 2002; Salaneck et al., 2000). In humans, single nucleotide polymorphisms are found frequently for Y2 receptor mRNAs, including those in the coding exon (Hunt et al., 2011), but thus far there are no studies confirming effects of mutations upon function of the Y2 receptor. A common variant of the Y2R might be protective against obesity in humans (Lavebratt et al., 2006).

The large difference in response to nucleotide site antagonists for rat Y2 receptors in hypothalamic vs. other limbic or cortical areas (see Figure 7B) could mainly be related to differences in compartmentalization and / or coupling to transducers. The prevailing Y2 population in the parietal cortex (PAR) has large sensitivity to chaotropes (Figure 4B, 5B), a lower affinity than measured in other areas (Figure 4 and Table 4) and a relatively high sensitivity to GγS and GβS (Figure 7), and could be constituted mainly of Gi-coupling monomers. The density of the PAR Y2 sites is much below the corresponding Y1 value (Table 4).

1.5. Types and Carriers of Reactivity in the NPY System

The production and secretion of NPY in the mammal is apparently restricted to neural cells. This complements the apparent absence of a significant peptide YY (PYY) and pancreatic polypeptide (PP) production in the brain. Across mammalian species, the affinities of NPY and PYY for the same Y receptor subtype are generally similar (Parker and Balasubramaniam, 2008; Statnick et al., 1998). However, the Y1 and Y2 receptors have much higher affinity for NPY and PYY than the Y4 and Y5 receptors (see e.g. Table 1 in (Parker et al., 2011).

Across vertebrate species, the C-terminally amidated aromatic residues of Y peptides possess a large hydrophobic and benzene-ionic (Dougherty, 1996) interactivity, coupled with polarity of the amide group. This highly reactive C-terminus can serve for association with multiple receptors and other targets. The "free acid", C-terminally carboxylated, form of NPY has very low affinity for Y receptors (Li and Hexum, 1991). The enormous affinity difference of the free acid and amidated forms of Y peptides is probably instructive in the matter of coexistence of opioid and NPY receptors on the same cells (e.g. (Williams et al., 2011)), since discharges of enkephalins and endorphins, which have the free-acid C-termini, would not strongly affect the Y sites, and *vice versa*.

In NPY and related PYY the interactivity is enhanced by the strongly cationic 22-34 region also containing multiple large hydrophobic sidechains. In quite similar pancreatic polypeptide (PP) , the cationicity of the 24-36 zone is internally compensated by backfolding upon the anionic N-terminal half (Table 1), to form the 'PP-fold'(Blundell et al., 1981; Lerch et al., 2002), resulting in a low non-specific interactivity (Parker et al., 2011), lack of binding to the Y1 and Y2 receptors and a low affinity at the Y5 receptor (Parker et al., 2011), and a lesser ability to stimulate feeding (Kanatani et al., 2000). The quite variable sequences of pancreatic polypeptides (Table 1) and Y4 receptors (Larhammar, 1996; Larhammar and Salaneck, 2004) appear as an evolutionary experiment still in progress.

NPY associates strongly with vertebrate Y1 and Y2 receptors, and also has significant affinity for the Y5 receptor, for the mammalian organismic PP Y4/(Y5)-like receptor (Parker et al., 2001a; Parker et al., 2002b; Parker et al., 1998a), and for the chicken Y4 receptor (Lundell et al., 2002). The glucocorticoid-induced receptor (Sah et al., 2007) and prolactin-releasing peptide (PrlRP) receptor ((D'Ursi et al., 2002; Hinuma et al., 1998); see Table

1 for the structure of PrlRP) also bind NPY, with profiles somewhat resembling that of the Y2 receptor.

The Y1 agonists internalized with Y1 receptors are disposed via intracellular traffic of endosomal rafts, which accumulate in the perinuclear zone (Pheng et al., 2003; Zhang et al., 1994) and could even deliver agonists into the nucleus (Silva et al., 2005), to potentially influence transcription. This type of localization of internalized receptors is quite common for heptahelical GPCRs with basic peptidic agonists, including neurokinin-1 (Garland et al., 1994), gastrin-releasing peptide (Grady et al., 1995), chemokine (Signoret et al., 2000), ghrelin (Camina et al., 2004) and apelin (Lee et al., 2004) receptors. However, all of the above receptors also possess large ionic residue-rich intracellular tracts which can function as the "nuclear localization" trafficking motifs (Lee et al., 2004). The perinuclear zone could be a staging ground for exchange traffic of peptides and proteins with multi-ionic motifs. The ultimate proteolytic disposal of Y1 agonists is strongly dependent on receptor internalization (Parker et al., 2007b) and proteolysis of the agonist may also occur in the perinuclear compartment (Oksche et al., 2000).

NPY has abundant less-specific interactions, usually lumped in the category of the "non-specific", linearly increasing non-saturating binding (see e. g. (Parker et al., 2011) on this subject). This is connected to the distinct segregation of ionic residues, with the 5-18 and 19-36 sections containing, respectively, only acidic and basic ionizable sidechains (Table 1). These homoionic zippers are candidates for a direct activation of transducers and effectors. (In mammalian pancreatic polypeptides, the homoacidic part includes the entire N-terminal half of the molecule (Table 1); this could however be internally neutralized by back-folding (Blundell et al., 1981) of the homobasic section, the "PP-fold"). The acidic 5-18 stretch of NPY and PYY is mainly random-coiled, while the homobasic 19-36 part is (in phospholipid micelles) largely helical (see (Lerch et al., 2005) and pdb file 1TZ4). The 5-18 stretch should be responsible for most of the large non-specific binding of NPY(Parker et al., 2011). This binding could include association with a number of broad-range physical targets of NPY. However, with targets other than Y receptors, the same would apply even more to the 19-36 homobasic segment of NPY (see the next section).

The non-specific reactivity of NPY can be enhanced by preferential orientation of the homoacidic random-coiled / unstructured 5-16 peptide toward the aqueous phase, by analogy withY5 agonists (Bettio et al., 2002). This peptide sector differs considerably between NPY and PYY (Table 1), and could be responsible for the large difference in the non-specific binding between the two peptides (Parker et al., 2011). The C-terminal homobasic 19-36 sector shows only conservative differences between NPY and PYY (Table 1).

The feeding stimulus elicited by NPY should relate to glucose depletion due to cAMP reduction via NPY-stimulated Gi/o transduction (Noll et al., 1996). Glucose-sensitive neurons in the arcuate nucleus contain NPY (Muroya et al., 1999). A selective Y1 agonist (D-Arg25)NPY decreases plasma glucose in the rat in a delayed way after a brief central infusion (Gao et al., 2004). This corresponds with our finding of an extensive forebrain distribution of NPY injected into the third ventricle, with substantial presence of NPY in hippocampal and neocortical areas (figures 1 and 2). The appetite-promoting NPY neurons are directly inhibited by glucose (Burdakov et al., 2005). Culturing in a high glucose medium decreases NPY expression in neuronal cell lines and hypothalamic explants (Lee et al., 2005). High glucose reduces the NPY mRNA and increases the CART mRNA in rat insulinoma cell cultures (Arumugam et al., 2007).

1.6. Histological and Cellular Expression of Y1 and Y2 Receptors in Mammalian Brain

For histology of NPY receptors in rat brain the reader should also consult the article by Masliukov and Nozdrachev in this book. The Y1, Y2 and Y5 receptors are located on the same chromosome band in man (4q31) and human Y1R and Y5R even may share promoters (Herzog et al., 1997). The Y1 and Y5 genes are on the same chromosome also in the mouse (8B3), but located about 10 kilobases apart (Nakamura et al., 1997). A co-expression of the Y1 and Y5 receptors in the same cells is therefore to be expected. However, the Y1R and Y2R genes are on different chromosomes in the mouse, and in the rat the two receptors are regulated by sex steroids in a widely different fashion (Parker et al., 1996b). The presynaptic Y2R of the rat brain (Stanic et al., 2011) however does get co-expressed with the postsynaptic Y1R upon axon damage (Landry et al., 2000).

The Y1 and Y2 receptors have not been shown to antagonize in feeding of non-mammalian species, and the following consideration will be confined to mammals that were studied with respect to involvement of NPY in the regulation of feeding. In the first decade after discovery of the Y1 and the Y2 receptors (see (Wahlestedt et al., 1990) for an early summary), the consensus was largely in favor of preferential, and even exclusive, distribution of the two receptors to different brain areas. For example, the hippocampal tissue was singled out as expressing only the Y2 receptor. It eventually became apparent that both receptors are significantly expressed in most forebrain areas (Aicher et al., 1991; Dumont et al., 1993; Larsen et al., 1993; Parker et al., 1996a), as is also clear from Figure 4 and Table 4 in this chapter. The regional densities and types of receptors could be quite different across mammalian species. The human brain could have more of the Y2 subtype compared to the rat brain (Caberlotto et al., 2000; Caberlotto et al., 1997; Statnick et al., 1997), but that can also be influenced by transductional and autolytic changes due to the much delayed *post mortem* human brain tissue sampling and a different stability of the two receptors (see the section 2.7 in this chapter). In the rat, limbic areas are however clearly enriched in the Y2 subtype, and cortical areas contain most of the Y1 subtype (Table 4).

1.7. Association of NPY Receptors with Transducers and Effectors

The G-protein coupling receptors (GPCRs) expressed in epithelial-type cell lines usually can be isolated as stable dimers coupled to heterotrimers of G-proteins (see especially (Baneres and Parello, 2003)). These hetero-pentamers contain a majority of Y receptors in CHO cell expressions and in rabbit kidney cortex (Estes et al., 2008), but in G-protein rich forebrain tissue most of the receptors are detected as Gα-coupled monomers (Parker et al., 2008a). The dimers within heteropentamers are readily converted to Gα-associated monomers by exposure to agonists in the presence of divalent cations (Parker et al., 2008a). These monomers should be the principal transductionally active form of GPCRs (Arcemisbehere et al., 2011; Whorton et al., 2008). However, the βγ subunit complex, coupling to both receptors and effectors (Biddlecome et al., 1996; Ford et al., 1998; Mukhopadhyay and Ross, 1999; Schillo et al., 2004), could mainly link to the protomer that is weakly associated with the α subunit. This may help explain the ubiquitously detected one dimer / one heterotrimer constitution of stored GPCR complexes with G-proteins. With GPCR heteropentamers, association with protein kinases and protein phosphatases (which is known for most GPCRs) may constitute an obligatory step in the transduction to a G-protein by the same or the neighboring protomer, but this needs further clarification.

The dimers of GPCRs form mainly by hydrophobic / H bonding interaction of bulky, especially aromatic, residues (Leu, Tyr, Trp, Phe) in transmembrane (tm) domains. Since the heptahelical transmembrane bundle is fairly symmetrical ((Palczewski et al., 2000); see also (Kim and Jacobson, 2006)), any of the helices can be used in dimerization, and indeed most have been identified with dimers (see (Filizola et al., 2006; Kim and Jacobson, 2006; Kota et al., 2006) and the review by (Parker et al., 2009)). The quiescent, non-activated forms of many GPCRs could be heteropentamers of receptor dimers and G-protein αβγ heterotrimers (Baneres and Parello, 2003). This is found consistently for the Y1, Y2 and Y4 receptors expressed in epithelial-type cells (Estes et al., 2008; Parker et al., 2008b; Parker et al., 2007b), and should also apply to Y5 homodimers (Dinger et al., 2003). GPCRs could form both homo- and heterodimers by coupling of transmembrane domains. Given the close association of Y1 and Y5 genes on the human chromosome 4 band q31(4q31) and similarity of promoters (Herzog et al., 1997), one would expect co-expression and then a significant hetero-dimerization in the ER due to similar spatiotemporal elements of mRNA transcription and protein synthesis. Indeed there is evidence for heterodimers of Y1 and Y5 receptors (Gehlert et al., 2007). However, the Y4 receptor does not readily heterodimerize with other Y receptors, or with opioid δ ad μ receptors (Berglund et al., 2003). Thus far, there also is no evidence for physiologic heterodimers of the Y2 receptor. The transmembrane domain-linked oligomerization of NPY receptors with other membrane-resident proteins is of course also possible, and this also has not been explored. The heteropentamers of Y2 receptors are largely compartmentalized in CHO cells (Parker et al., 2007c); however, the same apparently also applies to the Y2 monomers in rat hypothalamus (see (Parker and Balasubramaniam, 2008) and figures 7B and 8C).

The natively expressed Y2 receptors in brain and kidney and the cloned Y2 receptors expressed in CHO cells are detected as two populations of similar size but with 20-40-fold affinity difference (Parker et al., 1996a; Parker et al., 1998a; Parker et al., 2008a) (see Table 4 for rat brain Y2 receptors). After labeling in brain particulates with $[^{125}I]hPYY(3-36)$, both affinity populations are solubilized mainly as 80-100 kDa material with not more than 40% reactivity with Gi and Gq α subunit antibodies (Parker et al., 2008a). Since the Y2 receptor poorly couples with Go subunits (Freitag et al., 1995) and is not known to significantly interact with Gs type subunits, the residual material could represent complexes of the Y2 receptor with ~50 kDa partners, which can include especially the Gβγ dimers.

The activation of commonly employed transducers by different receptors could show, in an identical cellular milieu, wide variations dependent on subtypes and expression levels. As an example, the Gi-transduced inhibition of α-MSH release by NPY and dopamine shows much longer effect of NPY (Zhang et al., 2006). The difference could be due to use of different epitopes for the coupling of the Y1 and the D2 receptor to Gi α subunits. NPY can accumulate in the bilayer, enabling a protracted activity. The Y1 receptor could prefer the Gi2 or Gi3 subunit (Michel, 1998), and the D2 receptor may work best with the Gi1 subunit (Grunewald et al., 1996). Also, in the same cellular environment, the levels of different Gi subunit subtypes and their preferred subtypes of adenylyl cyclases could differ significantly.

1.8. NPY Uses NPY Receptors, but May Also Work Directly with Effectors

The feeding stimulation by NPY should mainly proceed through Y1 and Y5 receptors, although central Y2 receptors could also participate. At high levels of circulating NPY, this may even induce inhibition by leptin (Serradeil-Le Gal et al., 2000) and insulin (Moltz and McDonald, 1985; van den Hoek et al., 2004). Upon a third ventricle injection of NPY, the hypothalamic Y1 sites are the main targets (Stanley and Leibowitz, 1985), but one should not discount a role for other limbic sites (see figures 1 and 2) and for the very abundant cortical Y1 sites ((Aakerlund et al., 1990; Aicher et al., 1991) and Table 4). These receptors are likely to have a long-term contact with NPY injected into the ventricles, or with any synaptically released NPY, as will be shown in Section 2.1. The brainstem / medullar vagal-loop feeding-suppressing Y2 receptors (Batterham et al., 2002) are certainly also NPY targets, but these areas are far from the principal zones of NPY release. However, brainstem sites should receive significant inputs of systemic PYY(1-36) and PYY(3-36) (Hernandez et al., 1994; Nonaka et al., 2003; Dumont et al., 2007). The affinity of NPY at the Y2 receptor is very high and similar to that of PYY and of N-terminally clipped 34-peptides NPY(3-36) and PYY(3-36) (see Table 2 in (Parker and Balasubramaniam, 2008)), and medullar receptors should be adequately saturated at peptide concentrations below 100 pM. Also, the Y2 receptors could represent the majority of Y receptors in the hypothalamus (Aicher et al., 1991).

NPY above 3 nM acts as a partial antagonist of the Y1R in CHO-Y1 cells and SK-N-MC cells (Parker et al., 2007a; Sah et al., 2005), which is not observed with PYY and PP even at concentrations larger by an order of magnitude (Figure 10A, B). Under the same conditions, NPY blocks the Y2 receptor less than 50%, and PYY has about the same efficacy (Figure 10C, D). NPY also has a strong competitive activity at orexin receptors (Kane et al., 2000), and even a moderate activity at chemokine receptors (Norenberg et al., 1995; Shamri et al., 2002) and cation transporters and channels (Prieto et al., 1997).

Table 2. Homoionic tracts in domains of voltage-gated cation channels, NPY and NPY receptors

Channel	Number of residues in domain(s)	Acidic tracts % domain's aa	Basic tracts % domain's aa	% domain's acidic aa in acidic tracts	% domain's basic aa in basic tracts	% ionic aa in acidic tracts	% ionic aa in basic tracts
Voltage sensors (S4 transmembrane helices) in channels							
CAC	20.1	0	77.7	0	100	0	36.4
SCN	21.1	0	73.6	0	100	0	37.3
KCN	21.9	0	67.9	0	99.6	0	38.5
Extracellular domains between the S5 and S6 pore helices in channels							
CAC	77.6	40.2	16.4	85.2	64.9	27.2	32.6
SCN	79.8	35.1	16.7	78.5	65.4	30.4	39
Intracellular domains after the S6 pore helices in channels							
CAC	572	24.9	34.9	78.3	83.8	38.7	31
SCN	225	39.5	23.3	83.5	73.2	34.1	38.5
KCN	233	29.1	34.7	78.8	82.9	36.6	35.1
Homoionic tracts of human/rat neuropeptide Y							
NPY	36	30.6	47.2	100	83.3	45.5	29.4
Pooled intracellular domains of human NPY receptors							
Y1 receptor	117	16.2	43.6	64.3	89.3	47.4	40.6
Y2 receptor	106	4.72	43.	25	88.9	40	42.4
Y4 receptor	105	3.81	45.7	50	86.4	75	28
Y5 receptor	178	5.06	51.1	46.2	88.1	66.7	35.2

CAC = α subunits of 9 human voltage-gated Ca2+ channels; SCN = α subunits of 9 human voltage-gated Na+ channels; KCN = α subunits of 33 mammalian voltage-gated K+ channels. The acidic and basic tracts are segments containing two or more respective ionic sidechains (Parker et al., 2012a).

Effects of in-membrane NPY(McLean et al., 1990; Zerbe et al., 2006) could be important here, including differential association with receptor-associated transducers and effectors.

Homoacidic tracts with ≥10 acidic sidechains in several classes of human cytosolic effectors and shuttle proteins which are known to interact with GPCRs are shown in Table 3.

Table 3. Large homoacidic zippers in human cytosolic proteins that could bind to intracellular domains of GPCRs

Group [number of proteins]	number of tracts	number of acidic residues	number of all residues	% acidic in tracts
Shuttle [12]	21	16.6 ± 1.6	28.9 ± 1.9	58.9 ± 4.3
Ser/Thr protein kinases [18]	24	12.4 ± 0.53	35.6 ± 2.7	38.2 ± 2.5
Ser/Thr protein phosphatases [8]	16	12.4 ± 0.49	25.3 ± 3.5	60.1 ± 6.9
Non-receptor Tyr protein kinases [5]	5	10.8 ± 0.58	31.6 ± 5.5	42.4 ± 12
Phospholipases C-β [4]	4	15.3 ± 2.3	32.0 ± 7.5	53.8 ± 11

The averages for all homoacidic tracts with 10 or more acidic sidechains in the respective groups are shown ± 1 s.e.m. The number of acidic residues is the average of acidic sidechains per tract.

The number of all residues is the average of all aa per tract. The % acidic in tracts is the average % of acidic aa per tract (% acidic density).

These tracts obviously are candidates for association with the highly homobasic intracellular segments of Y receptors (Table 2).

1.9. Effects of NPY and Other Y Peptides on Channels and Transporters

The Y peptides, including NPY, have been implicated in activation and inhibition of numerous channels and transporters, in most cases with participation of the corresponding specific receptors and through activation of specific transducers. In the case of rat or human pancreatic polypeptide (PP) there is strong dependence of binding to four mammalian Y4-type receptors upon $[R^+]$ in the medium, with Na^+ being much more potent regulator than K^+. This binding has a high sensitivity to N5-substituted, but not to C2-substituted amilorides (Parker et al., 2002b), indicating association of the receptor with amiloride-sensitive sodium transporter(s)/exchanger(s). Human, rat and guinea-pig Y4 receptors expressed in CHO cells and the rabbit/rat (kidney, forebrain and pituitary)Y4/(Y5) PP receptors (Parker et al., 2000; Parker et al.,

1999) show a low sensitivity to GTP and GγS and not more than 40% sensitivity to pertussis toxin (PTX), and the sensitivity of PP binding to N5-amilorides is retained after Gi/o inactivation by PTX (M.S Parker and S. L. Parker, in preparation), confirming association with Na^+ transporter(s). The Y4 receptor appears to have a major association with transporter(s). This example should serve as a reminder that heptahelical receptors, including the Y receptors, should not be necessarily expected to transduce through pre-coupled GTPases. A precoupling to channels and transporters (in the case of the Y4 receptor possibly through the highly basic third intracellular loop (see (Parker et al., 2002b))) could be effected at the level of receptor completion in the ER (Hilairet et al., 2001; Yasuda et al., 2009), or at an endosomal stage.

A brief review of effects of NPY on the activity of calcium and potassium channels points to involvement of NPY receptors and possibly also to direct association of NPY or related peptides with the channels. NPY is frequently reported to activate inverse-rectifier potassium channels (Brown et al., 1995; Prieto et al., 1997; Sosulina et al., 2008; Sun et al., 2001; Zylbergold et al., 2010). This is usually ascribed to NPY-triggered Y1 transduction to Gi α subunit, which is presumed to directly regulate the KIR/GIRK channels (Brown et al., 1995; Zylbergold et al., 2010). However, this activation appears to have a G-protein independent component (Prieto et al., 1997), and could also link to direct binding of homoionic zippers of NPY to tracts within channels.

NPY enhancement of the release of pituitary luteinizing hormone could be mainly due to interaction with GnRH receptor, rather than the scant pituitary NPY receptors (Crowley et al., 1990; Parker et al., 1991). There is Y2 receptor-linked activation of calcium influx in CHP-234 neuroblastoma cells, which is PTX-insensitive (Lynch et al., 1994), and trophic actions of NPY in cardiomyocytes also could be linked to the Y2R, and not be PTX-sensitive (Protas et al., 2003); however, there is inhibition by NPY of the N-type Ca^{2+} channel in SH-SY5Y neuroblastoma (McDonald et al., 1995). The adrenal nAChR inhibition by NPY that was ascribed to the hypothetical y3 receptor not affected by PYY (Norenberg et al., 1995) may relate to the essential acidic residues in the N-terminal extracellular domains of the α subunits of nAChRs (Sala et al., 2005). There is mobilization of Ca^{2+} in neuroblastoma cells by PYY(3-36) (Connor et al., 1997). NPY causes a short-term inhibition of N and P/Q-Ca currents (voltage-independent, presumably through the Y2R) and a longer activation of GIRK (via the Y1R) in thalamic neurons (Sun et al., 2001).

The inhibition by NPY of the presynaptic Ca^{2+} influx in the retinal rods could be Y2-linked (D'Angelo and Brecha, 2004). The inhibition of Ca^{2+} currents and activation of GIRK by PYY(3-36) in NPY neurons should be involved in anorexic activity of the peptide (Acuna-Goycolea and van den Pol, 2005).

In cerebral cortex neurons, inhibition by NPY of N- and P-type calcium channels and glutamate release might depend on the Y1R(Wang, 2005), and the long-lasting inhibition of ICa^{2+} by NPY in melanotropes could also be via the Y1R (Zhang et al., 2006).

A similar inhibition by NPY or by neuropeptide FF of the N-type calcium channel could proceed via Gi α subunit coupling to Y2 and NPFF2 receptors (Mollereau et al., 2007), but it might also be due to action of peptides on ionic tracts of the channel.

The above very diverse responses seem to collectively indicate a two-component, receptor-linked and direct, actions of NPY.

Tables 2 and 3 summarize the homoionic parameters of channels, NPY itself, NPY receptors and effectors known to interact with GPCRs. These subjects are considered in the General Discussion section.

1.10. Blockade of NPY Receptors by Excess NPY May Lead to Inverse Agonism

Accumulation of NPY in the bilayer may relate to the observed blockade of Y1 receptors in CHO and SK-N-MC cells by excess exogenous NPY (Parker et al., 2007a; Sah et al., 2005). This blockade could mainly depend on interactions of the 10-16 EDAPAED segment, and is not observed with PYY (with 10-16 EDASPEEL segment; Table 1), which has about 2.5-fold lower non-specific binding (Parker et al., 2011). It should be noted that NPY does dimerize even in free solution (see (Chatenet et al., 2011)), and the dimerization could be promoted in lipid-rich environments. This may increase the retention and accumulation of NPY in the bilayer, and could be related to blocking of the Y1R traffic by high concentrations of NPY (Parker et al., 2007a).

The persistent blockade of the CHO-Y1 (or SK-N-MC Y1) receptor, but not of the CHO-Y2 receptor, by high concentrations of NPY ((Parker et al., 2007a; Sah et al., 2005); Figure 10), is not observed with PYY, and could relate to a stable association with currently unidentified 80-160 kDa partners, and contribute to a protean inverse agonism. PYY has a much lower stable association with targets other than Y1 receptors ((Parker et al., 2011; Parker et al., 2007a; Sah et al., 2005) and Figure 11), and similar is observed for PYY(3-36).

2. Experimental

Much of the experimental work presented here was reported on the Society for Neuroscience meetings from 1994-2008. The data presented in figures 1-7 and 9 were not previously published in any form. The results in Figure 8B and C, Figure 10 and Figure 11 are adapted from previous publications. While some of the data overlap prior publications by ourselves and other authors, to our knowledge there is no comprehensive comparison of the Y1 and the Y2 receptors in cortical and limbic areas of rat brain that covers the agonist binding, chaotrope sensitivity and association with transducers. That type of comparison is also essential in interpretation of the current knowledge concerning NPY physiology and pharmacology.

2.1. NPY Distribution and Fate upon Injection in the Third Ventricle

All distances posterior to bregma and lateral are in millimeters, and are based on the atlas of Paxinos and Watson (second edition, 1986). Tissues were pooled from 10-20 animals for each area and processed as described in (Parker et al., 1998b). The following list identifies the brain areas used and the approximate stereotaxic coordinates. The numbers in parentheses indicate start and end of the respective cuts, in mm behind bregma. All cuts are bilateral.

POA	The circumventricular preoptic hypothalamus (0.3 -1.2)
MBA	The circumventricular medial basal hypothalamus (1.8 -3)
POST	Posterior hypothalamus (3-5)
SON	Supraoptic nuclei (0.9 - 2.1)
PVN	Paraventricular nuclei, 1.2 - 2.1
AHA	Anterior hypothalamus (POA + MBA)
PAR	The parietal cortex area 1 (0.6 - 3)
PIR	The piriform cortex area (0.6 - 3)
HIPP	The anterior hippocampus (1.8 - 3)

Cortical areas are outside the major waterways of the forebrain, which are in the limbic system. However, neocortex evolves by radial unit migration from the proto-ventricle (Rakic, 1988) and retains direct csf connections to ventricular space. After injection in the 3V there is quite significant penetration of the injected peptide to PIR and PAR within 20 min (Figure 1).

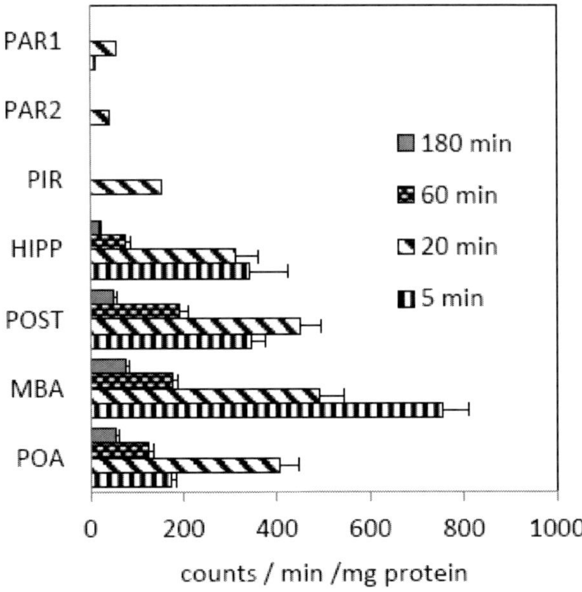

Figure 1. Recovery of NPY-derived radioactivity in rat brain areas at 5-180 min following injection of [^{125}I] NPY into the hypothalamic aspect of the third ventricle. The cannula placement was at 1.2 mm behind bregma and the injected amount of NPY was 100 fmoles (about 0.43 ng). Definition of the brain areas is given in section 2.1.

The anterior hippocampus obviously receives NPY via the dorsal third ventricle (which directly contacts the dentate gyrus) and as a consequence is quite significantly labeled even at 5 min post-injection (Figure 1). The large distribution to hippocampus as early as 5 min postinjection attests to a fast flow from the third to the lateral ventricles. At 20 min postinjection there is considerable distribution to piriform cortex, and a detectable flow to parietal cortex, i. e. to areas that are outside the limbic brain and far from major cisterns of the csf.

There is non-saturable transport of intact NPY through the cerebral ependymal blood-brain barrier (BBB) in both directions ((Kastin and Akerstrom, 1999); see also Figure 3). However, while numerous studies examined effects of intracerebrally administered NPY upon food intake, the intracerebral distribution and fate of the injected peptide apparently were not studied. We present some basic information in this regard (figures 1-3).

Involvement of cortical NPY receptors in the control of feeding could be larger than is commonly perceived. In the rat, cortical NPY receptors are abundant, and especially the Y1 subtype (Dumont et al., 1993). These receptors were not studied extensively in connection to feeding, although the lateral ventricle injection of NPY and PYY is known to induce feeding (Corp et al., 2001b). The feeding stimulation by a bolus injection of high doses of NPY or PYY (0.25 -1 nmol) into the third ventricle results in a vigorous feeding response within 2 min, which lasts up to 60 min. Significant amounts of NPY are found even 3 hours after the third ventricle injection (Figure 2). The injected amount (100 fmol, or about 0.43 ng of NPY) could be in the range of synaptic discharges of NPY. Non-degraded NPY is 75% of total recovery at 5 min, and 62-65% at 20 min after injection. The size distribution observed at 20 min should be close to an equilibrium csf profile, since the degradation products do not accumulate in the csf. A significant presence of NPY(3-36) and of the C-terminal hexanucleotide ITRQRY is detected at equilibrium. Free tyrosine elutes at the total volume (V_t). Undegraded NPY is detected in the brain even 180 min after injection of this small dose. This corresponds with the known long-term feeding stimulation upon a bolus injection of NPY. The above findings are compatible with a slow disposal due mainly to internalization and intracellular processing. This is expected from the findings of (Kastin and Akerstrom, 1999) and (Frerker et al., 2007) about processing of Y peptides in the rodent brain. In serum there is a large presence of short C-terminal peptides (Figure 3), which apparently accumulate due to low affinity for the systemic Y2 autoreceptors; the K_{diss} of NPY(26-36) is 115 nM (Parker and Balasubramaniam, 2008), and that of TRQRY.NH2 is 2.1 μM (Leban et al., 1995). [^{125}I]Tyr in serum is very low at 180 min, reflecting an expected proper clearance. NPY(1-30) (Frerker et al., 2007) and NPY(3-35) (Abid et al., 2009) are present, but not significantly contributing to the above profiles, since (based on digestion with trypsin) ~90% of the ^{125}I tracer is in the C-terminal Tyr36.

Feeding regulation involves voluntary and even cogitative components, especially in humans. Studies of feeding regulation by NPY are rather restricted to hypothalamic sources of the peptide, such as arcuate nucleus (Jhanwar-Uniyal et al., 1993) or paraventricular nucleus (Sahu et al., 1992), and hypothalamic targets, especially the nuclei surrounding the hypothalamic aspect of the third cerebral ventricle. However, administration of NPY into the lateral ventricle is quite efficient in stimulating feeding in the mammal (Miner et al., 1989).

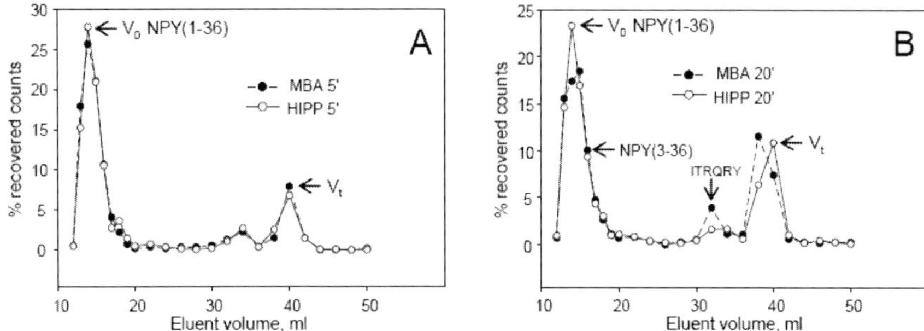

Figure 2. NPY is not rapidly degraded in rat hypothalamus or hippocampus after microinjection into the hypothalamic aspect of the third ventricle. The amount of [^{125}I]hNPY injected was 100 fmoles (about 0.43 ng) . (A) Profiles of MBA and HIPP extracts at 5 min after the injection. (B) Profiles at 20 min after the injection. The cannulas were placed at 1.2 mm behind bregma. Chromatography of the 100 mM HCl-extracted material was done on 30-ml columns of Bio-Gel P-4 in 1M CH$_3$COOH-0.05% BSA.

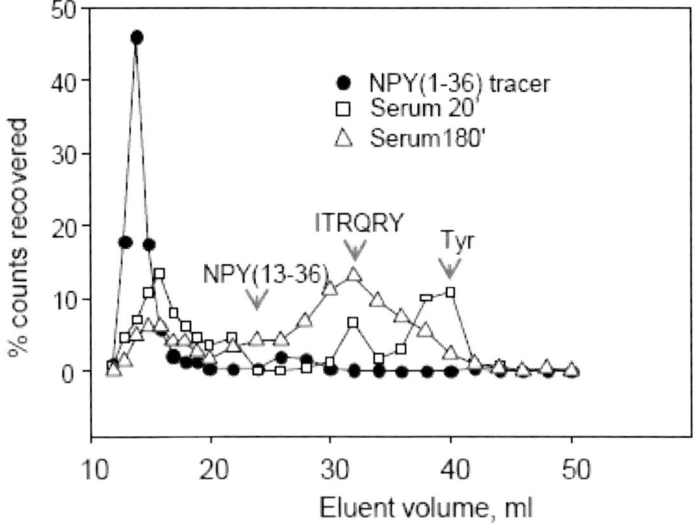

Figure 3. NPY passing the blood-brain barrier following icv injection is found as up to 50% degraded in serum, but there is a significant undegraded component even at 3 hours after the injection. Extraction and Bio-Gel P-4 chromatography were performed as in Figure 2.

Rat hypothalamus contains NPY in more than fivefold excess over cortex (Allen et al., 1983), but cortical Y1 receptors in the rat outnumber the hypothalamic by more than one order of magnitude by numbers, and at least twofold by density (Table 4). Any NPY released in the periventricular hypothalamus is going to be quickly distributed to other limbic areas, and even to cortical, and shall persist in the csf for considerable periods (Figure 1). The high affinity of Y1 receptors would assure an extended stimulation even at locations remote from the site of synaptic release (or injection) of NPY. With high-affinity K_{diss} for the Y1 receptor of about 12 pM (see Figure 4 and Table 4), a sufficient saturation should be achieved at 100 pM, i.e. by administering 0.5 µg / kg brain of NPY (0.8-1 ng per an adult rat brain). Typical dosage ranges in feeding studies are from 20 -1000 ng per rat brain, highly likely to saturate all limbic as well as cortical Y1 receptors by way of the cerebrospinal fluid (csf) flow. NPY clearance is not fast, with undegraded NPY being found at 3 h after the icv injection of only 100 fmoles of NPY even in rat blood (Figure 3). There is >60% undegraded NPY in hippocampus 20 min after icv injection of 100 femtomoles of NPY (Figure 2) (providing an equilibrated limbic concentration of about 100 pM).

In the rat, Y1 receptors (NPY targets for activation of the adenylyl cyclase-inhibitory Gi/o α subunits that lower the production of cAMP) are found principally in cortical areas, which have some efferents to the vagal/NTS circuitry (Hurley-Gius and Neafsey, 1986). A decrease of glucose levels in these neurons is likely to produce feeding stimuli. NPY injected into the third cerebral ventricle (icv) is present in cortical areas already by 20 min (Figure 1).

The injected or released NPY is not rapidly degraded and cleared from the csf, and should be present there for extended periods even after injection of small dosages.

2.2. The Binding of Y1 and Y2 Subtype-Selective Agonists to the Particulate Y Receptors

Numerous studies have provided detailed maps of Y1 and Y2 receptor distribution and abundance in rat brain (Aicher et al., 1991; Dumont et al., 1993; Dumont et al., 1995; Larsen et al., 1993). We have added to this by suppressing the possible Y4 and Y5 contributions by masking with Y4-selective pancreatic polypeptide and the Y4/Y5-neutral Y1 antagonist BIBP-3226.

Table 4. Binding parameters for Y1 and Y2 sites in particulates from rat forebrain areas

Area	K_{diss} high, pM	K_{diss} low, pM	% binding K_{diss} high	C_{max}	% high affinity
Y1 binding					
HIPP [6]	11.8 ± 1.2	296 ± 29	25.7	60.3 ± 9.3	10.8
MBA [6]	4.97 ± 0.4	114 ± 41	23.2	35.1± 4.3	18
PAR [13]	8.52 ± 1.6	106 ± 34	12.5	296 ± 50	6.71
PIR [8]	30.7 ± 11	310 ± 68	18.6	210 ± 23	3.31
POA [3]	9.43 ± 1.4	603 ± 169	9.4	65.8 ± 13	4.10
Y2 binding					
HIPP [12]	8.57 ± 1.9	142 ± 22	39.0	15.3 ± 1.1	57.1
MBA [14]	7.54 ± 1.8	73.1 ± 25	32.8	23.4 ± 8.3	58.1
PAR [7]	26.9 ± 3.2	502 ± 96	5.08	11.1 ± 2.7	27.7
PIR [10]	6.07 ± 1.2	96.9 ± 15	19.8	25.5 ± 4.9	23.0
POA [9]	8.03 ± 1.5	142 ± 47	38.0	29.3 ± 7.5	39.8

Abbreviation: C_{max}, the maximum binding (fmol/mg particle protein) constrained in the LIGAND program (Munson and Rodbard, 1980) to the K_{diss} values shown below.

The number of binding runs is shown in the brackets after the area labels. For Y1 labeling, [^{125}I] (Pro34)hPYY was competed by 10-15 different concentrations of unlabeled (Pro34)hPYY in the range of 0.01-20 nM. For Y2 labeling, [^{125}I] hPYY(3-36) was competed by 10-15 different concentrations of unlabeled hPYY(3-36) in the range of 0.003 − 20 nM. Addition of 1 nM of hPP and hPYY(3-36) in Y1 assays or (Pro34)hPYY in Y2 assays did not significantly change the binding parameters (see Figure 4). For the constrained Y1 fitting, K_{diss} high was set to 12 pM, and K_{diss} low to 300 pM, and for the Y2 fitting K_{diss} high was set to 8 pM, and K_{diss} low to 120 pM. All constrained K_{diss} values are close to averages of the respective parameters from unconstrained biexponential fits (shown in columns 2 and 3).

Also, the Y1 -selective agonist binding was performed at saturating concentration (1 nM) of a Y2-selective agonist, and the Y2-selective binding was similarly done with 1 nM Y1 masking. These profiles are shown in Figure 4. As a novel finding, the parietal Y2 sites were found to have low density and affinity compared to other areas.

Rat forebrain abundantly expresses the Y1 and Y2 receptors (Aicher et al., 1991; Dumont et al., 1993; Larsen et al., 1993). Expression of the Y5 receptor in the brain appears to be low (Dumont et al., 1998; Statnick et al., 1998), and similar applies to the Y4 receptor (Campbell et al., 2003).

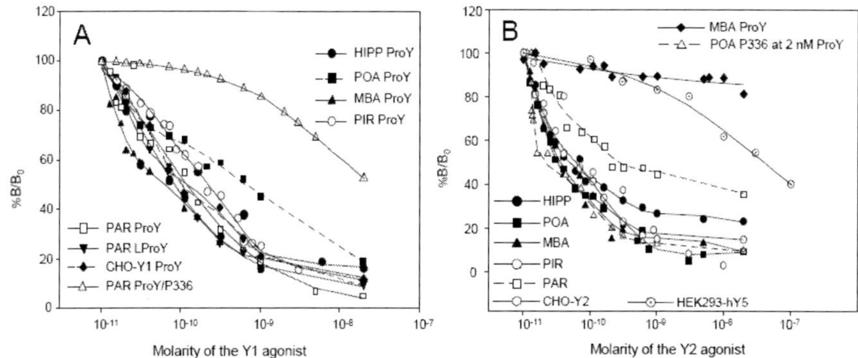

A   The binding of Y1-selecive agonists (Pro34)hPYY ("ProY") and (Leu31,Pro34)hPYY ("LProY") competed by 8-15 molarities of the same unlabeled peptides in the range of 1x10^{-12} to 3x 10^{-8} M. Also shown is the profile for competition of ProY binding by hPYY(3-36) ("P336").

B The binding of Y2-selective agonist hPYY(3-36) ("P336") competed by 8-15 molarities of unlabeled P336 in the range of 3x10^{-12} to 3x10^{-8} M. Also shown are the profiles for competitions of the MBA binding of P336 by ProY, of the POA binding of P336 by itself at 2 nM ProY, and of P336 binding to hY5 receptor expressed in HEK-293 cells.

Figure 4. Competitive binding of subtype-selective Y receptor agonists to particulates from rat brain areas and other sources. All profiles are averages of at least three binding experiments.

The Y1-receptor group agonists such as (Pro34)PYY and (Leu31,Pro34)PYY combined with Y4-selective rat or the (very similar) human pancreatic polypeptide and the Y5-unaffecting Y1 antagonists such as BIBP3226 (Dumont et al., 1998) could then produce reliable estimates of forebrain Y1 sites. Similarly, the Y2-selective agonist PYY(3-36) (Dumont et al., 1993) masked by the Y1/Y4/Y5 agonist (Pro34)PYY should allow appropriate quantitation of the forebrain Y2 sites. The Y1 profiles reported in this study were obtained using 25-50 pM [^{125}I](Pro34)hPYY or [^{125}I](Leu31,Pro34)hPYY at 1 nM rat PP or human PP, and 1 nM hPYY(3-36). The Y5 binding of these agonists was below 5% of the Y1 binding in any of the areas, as judged by the difference of the binding at 10 µM BIBP3226 and 100 nM (Leu31,Pro34)hPYY or (Pro34)hPYY. The Y2 profiles were obtained at 25-50 pM [^{125}I]hPYY(-36) , also masked by 1 nM (Pro34)hPYY and 1 nM hPP. The Y1-selective antagonist BIBP3226 was only active vs. the PYY(3-36) binding above 10 µM.

Figure 4A shows the binding of Y1-selective agonist (Pro34)hPYY to brain area and CHO-Y1 particulates. The high-affinity Y1 binding K$_{diss}$ values range from 5-12 pM in all areas except PIR. The low-affinity component is much more variable, probably related to differences in rates of activation and rundown across the areas. The Y1 receptor density is much higher in cortical areas compared to limbic. The limbic areas also show appreciably more Y1 sites than the Y2 (Table 4).

Figure 4B shows the binding of the mammalian Y2 receptor-selective agonist hPYY(3-36)

The high-affinity Y2 binding is uniform at 6-9 pM excepting PAR (27 pM). The density of PAR high-affinity sites is significantly below that in other areas. In HIPP and MBA, the high-affinity Y2 binding represents almost 60% of all binding, indicating a low rate of rundown; this applies to preoptic hypothalamus as well, but not to the cortical areas. The low rundown of hypothalamic Y2 receptors should be connected to a higher masking / compartmentalization which is obvious from effects of GβS (Figure 7B) and of preincubation with GγS (Figure 8C).

The low rat PAR Y2 density would reduce antagonism of Y1 activity by NPY-like peptides at these sites. This could also be of importance in human subjects. The human cortical Y1 and Y2 densities are however somewhat controversial due to problems with preferential *post mortem* autolysis of the Y1 receptor ((Caberlotto et al., 1997; Jacques et al., 1997); also see Figure 9). The Y1 mRNA is much more abundant than Y2 mRNA in the human cortex (Statnick et al., 1997). As we show for rat forebrain tissue in Figure 9, the autolytic Y1R changes are much more severe than those for the Y2 sites.

2.3. Inhibition of Agonist Binding to NPY Receptors by NaCl

Sensitivity of agonist binding to chaotropic agents is much higher for the Y1R compared to the Y2R (Parker et al., 1996a; Parker and Parker, 2000), and this applies to non-selective agonists as well, as shown in figures 5 and 6. The low-affinity Y2R binding in the parietal cortex again stands out as having much larger sensitivity to both ionic and non-ionic chaotropes than in other brain areas, or in CHO cell clonal expressions.

Ionic chaotropes are known to differentially affect the Y1 and Y2 binding to particulates from brain areas (Parker et al., 1996a), which relates chiefly to the two-prong attachment of agonist peptides to the Y1 receptor (Daniels et al., 1995). The Y1 binding is much more sensitive to monovalent cations, and this sensitivity is similar for two Y1-selective ligands, (Pro34)PYY and (Leu31,Pro34)PYY, and quite concordant across brain areas and Y1 expressions in cell lines (Figure 5A). These ligands are also Y4 agonists (Gehlert et al., 1997) and Y5 agonists. The Y4 receptor is expressed at low levels in the rodent brain (Campbell et al., 2003) and inclusion of 1 nM of pancreatic polypeptide does not alter the binding of Y1 agonists to rat brain area particulates. The binding at 10 μM Y1 antagonist BIBP3226 was, in any of the areas, not more than 5% above the non-specific binding (as defined at 100 nM of the subtype-selective agonists), confirming a low presence of Y5 sites (Dumont et al., 1998).

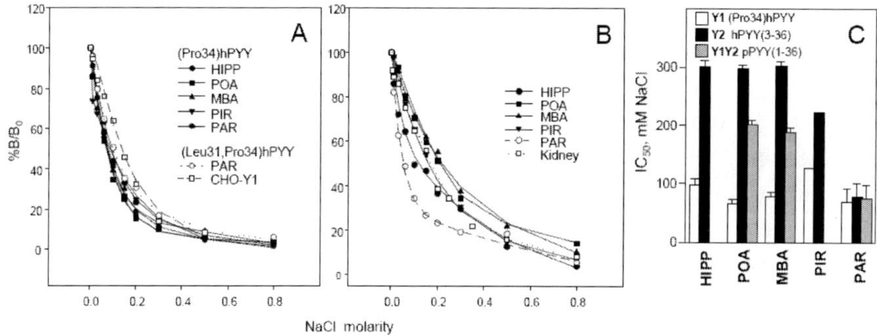

A   Inhibition of the binding of Y1-selective agonists (Pro34)hPYY or (Leu31,Pro34)hPYY. CHO-gpY1 = particulates from CHO cells expressing the guinea-pig Y1 receptor.
B   Inhibition of the binding of Y2-selective agonist hPYY(3-36). "Kidney" denotes particulates from rabbit kidney cortex. The binding to CHO-hY2 expression has sensitivity to NaCl very similar to that of the kidney Y2 receptor.
C   Values for half-inhibition by NaCl of the binding of subtype-selective ligands and of a non-selective ligand to particulates of the brain areas examined in detail in graphs A and B.

Figure 5. Inhibition by NaCl of the binding of NPY receptor subtype-selective agonists to particulates from rat brain areas and other sources. All profiles are averages of at least three binding experiments.

Inhibition of the Y1-selective agonist binding by NaCl is quite uniform across the brain areas and epithelial cell expressions, with ic50 values ranging from 70-130 mM. The Y2 binding (Figure 5B) is rather resistant to NaCl in hypothalamic areas, with ic50 values of about 0.3 M, but more sensitive in PIR, and has a high sensitivity similar to that of (Pro34)hPYY in PAR (see also Fig 5C). The non-selective agonist hPYY(1-36) shows in PAR particulates a high sensitivity to NaCl that is quite similar to that of subtype-selective agonists (Figure 5C). High mRNA levels contrasted by low expression of the human Y5 receptor (Statnick et al., 1998) may echo a co-transcription with the Y1R due to similarity of promoters and proximity of the Y1 and Y5 genes in the q31 band of human chromosome 4 (Herzog et al., 1997). However, in cell lines the Y5 receptor is much more difficult to stably express than the Y1, and the findings by Statnick and colleagues could reflect a similar situation with neural cells in vivo. A low retention of Y5 expressions could relate to the large third intracellular loop with massive homobasic segments that can promote retention in the ER as well as a constitutive internalization in response to low-affinity stimuli.

2.4. Inhibition of agonist binding to NPY receptors by urea

The sensitivity of agonist binding to Y receptors to non-ionic chaotropes was examined with urea, a prototypic agent of this type. As seen in Figure 6A, the inhibition of Y1-type binding is highly uniform across sources and quite linear with concentration of urea, suggesting multiple non-selective targets. The binding of (Pro34)hPYY and (Leu31,Pro34)hPYY tracers does not differ significantly in sensitivity to urea. With particulates from hypothalamic areas, the Y1 agonist binding sensitivity to urea is about twice larger than that for the Y2 agonist (Figure 6C). It should also be noted that the binding of [^{125}I]NPY to the CHO-Y1 receptor is much more resistant to urea than the binding of [^{125}I]PYY (K_i 0.96 and 0.6 M, respectively; (Sah et al., 2005)).

The inhibition of hPYY(3-36) binding (Figure 6B) however shows a much broader range of responses. The PAR Y2-like sites are by far the most sensitive. As with NaCl, the non-selective agonist hPYY(1-36) shows in PAR particulates a high sensitivity to urea quite similar to that of subtype-selective agonists (Figure 6C). However, no large change of sensitivity is detected in any particular range of urea concentration for any of the Y2 profiles, indicating (as with the sensitivity to NaCl) multiple overlapping contributions.

2.5. Sensitivity of the Binding of Subtype-Selective Agonists to Guanine Nucleotides

The rundown/inhibition of subtype-selective agonist binding by the irreversible GTPase nucleotide site activator guanosine-5'-(γ-thio)triphosphate (GγS) and the irreversible inhibitor guanosine-5'-(β-thio)diphosphate (GβS) were compared with particulates from brain areas and clonal expressions (Figure 7).

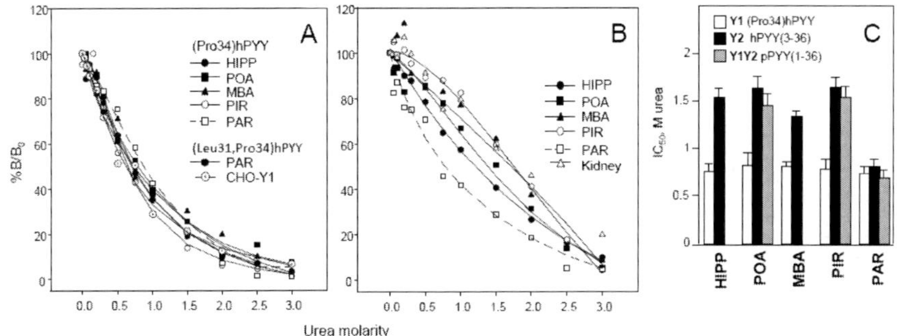

A The inhibition of Y1-type binding.

B The inhibition of Y2-type binding.

C Values for half-inhibition of the binding of subtype-selective ligands and of a non-selective ligand to particulates of the brain areas examined in detail in graphs A and B.

Figure 6. Inhibition by urea of the binding of subtype-selective agonists to particulates from rat brain areas and other sources. (Pro34)hPYY and (Leu31,Pro34)hPYY were used to label the Y1-type receptors, and hPYY(3-36) for the Y2-type. All profiles are averages of at least three binding experiments.

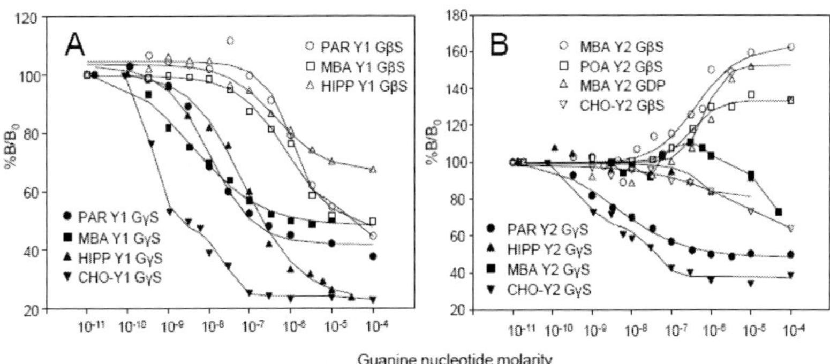

A The binding of Y1-selective agonist (Pro34)hPYY at 8-12 molarities of GγS or Gβs in the range of 1×10^{-10} to 1×10^{-4} M.

B The binding of Y2-selective agonist hPYY(3-36) at 8-12 molarities of GγS , GβS or GDP in the range of 1×10^{-10} to 1×10^{-4} M.

Figure 7. Sensitivity of agonist binding to Y1 and Y2 receptors to a blocking activator and a blocking inhibitor of G-protein nucleotide sites. All profiles represent at least three binding experiments.

This revealed much lower rundown by GγS for the Y2 receptor compared to the Y1, again excepting the parietal Y2 selective binding. In a striking difference with inhibition observed for the Y1-selective binding, there was a considerable activation by GβS, and also by GDP, of the Y2-selective agonist binding to particulates from circumventricular hypothalamic areas, indicating a difference in activation that could relate to masking/compartmentalization as well as to different transductional partners.

Among the principal types of G-protein α subunits, the Y1R is known to transduce via Gi and Go, and has a weaker activity at Gq subtypes. The Y2R is using Gi and Gq subunits and apparently is very weakly active with the Go type (Freitag et al., 1995), which is by far the most abundant α subunit type at least in cortical areas (Gierschik et al., 1986). Based on the low exchange rate of Gq subunits (Berstein et al., 1992) , we chose the rundown reduction of the binding of subtype-selective agonists by increasing concentrations of guanine nucleotides as the method that can provide a rough differentiation of Gi/o- and Gq-type coupling of the Y receptors in various brain areas. This approach should be more relevant than attempts to produce a dependable stimulation of inositol polyphosphate production or inhibition of cAMP formation by NPY or Y1-selective agonists, which in rat brain could be equivocal (e.g. (Michel et al., 1995)) due to abundant expression and activity of many other Gi/o-coupling GPCRs.

The binding of labeled agonist peptides is the usual method of quantitation of NPY receptors. This approach is complicated by dependence of the binding affinity on association with transducers. The heteropentameric complexes of Y receptor dimers and G-protein heterotrimers show much higher agonist binding affinity than Gα-associated monomers. Also, there are large differences between the Y1 and the Y2 receptor in sensitivity of agonist attachment to the status of the Gα nucleotide site. In all cell types examined (including kidney epithelia, piriform cortex, hypothalamic areas, CHO cells and SK-N cell lines), the Y2 binding is much less sensitive to GγS occupancy of the Gα nucleotides site than the Y1 binding. In hypothalamic areas the Y2 binding is even increased (probably due to unmasking; Figure 8C) by GγS, and especially by GβS (Figure 7B).

The responses of Y1 and Y2 agonist binding to GγS and GβS are quite different for rat cortical and limbic areas (Figure 7). The Y1 binding is driven down by GγS, with large differences across brain areas. As seen in Figure 7A and Table 5, the Y1 agonist binding in PAR, MBA and CHO-Y1particulates shows a component of similarly high sensitivity to GγS, which however in MBA is less than 20% of total GγS-sensitive binding, and does not exceed 50% of that material even with the CHO-Y1 expression. The sensitivity to GβS is at least two orders of magnitude less in PAR and MBA, and more than 1000-fold less in HIPP.

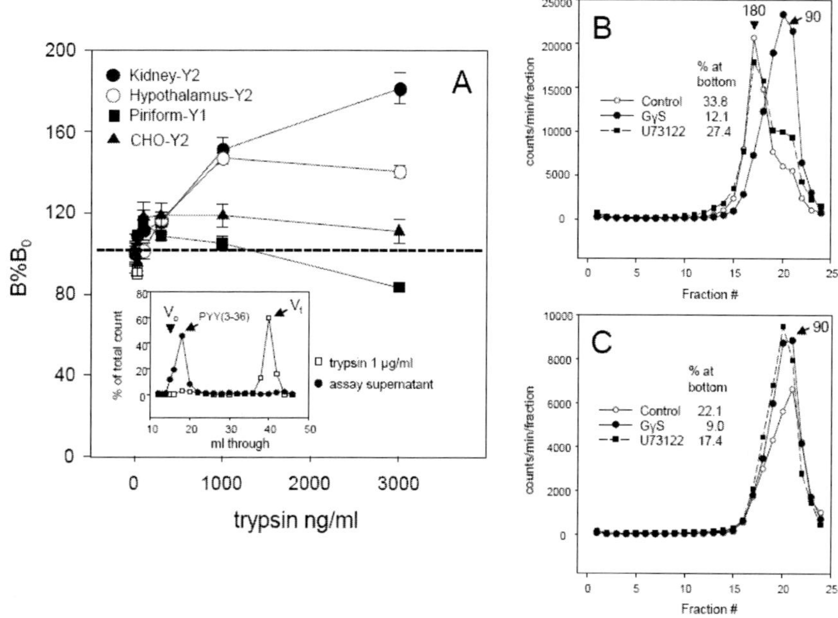

A Pretreatment with trypsin results in considerable unmasking of Y2 sites in particulates from kidney cortex and hypothalamus but not from CHO cells expressing human Y2 receptors. The dashed line indicates the binding for control without trypsin treatment. Trypsinization in 4 mM Ca^{2+} -10 mM Tris.HCl (pH 7.6) was done for 30 min at 37 °, followed by addition of 50 µg soybean trypsin inhibitor, sedimentation and washing prior to the binding assay. INSET: Bio-Gel P-4 profiles indicate complete preservation of tracer [^{125}I]hPYY(3-36) in the binding assays after trypsinization, as opposed to more than 90% degradation of the tracer by 1 µg/ml trypsin in 4 mM Ca^{2+} -10 mM Tris.HCl (pH 7.6) over 30 min at 37 °.

B, C Treatment with GγS, but not with phospholipase- C inhibitor U73122, disaggregates a large part of rabbit kidney (B) or rat anterior hypothalamus (C) Y2 sites that are not solubilized by digitonin and cholate (10 mM each). The particulates were labeled with [^{125}I]PYY(3-36), sedimented and then incubated for 15 min at 25 °C with the assay buffer (control), 10 µM GγS or 25 µM U73122 prior to solubilization and sedimentation for 6 h at 218,000 x g_{max}. Arrows indicate sedimentation of molecular weight (indicated in kDa) standards.

Figure 8. Unmasking of particulate Y2 receptors by trypsin and disaggregation by GγS.

        M. S. Parker, R. Sah, A. Balasubramaniam et al.

Table 5. The binding of Y1 and Y2-selective agonists challenged by GγS and GβS in particulates from rat brain areas

Area	Y1 GγS pK$_1$	% K$_1$	Y1 GγS pK$_2$	Y1 GβS pK$_1$	% K$_1$	Y1 GβS pK$_2$	GγS % change	GβS % change
PAR	8.05	32.4	3.59	6.05	36.4	3.56	-63.3	-55.6
MBA	8.18	19.6	4.31	6.10	34.9	3.14	-51.4	-52.2
HIPP	7.09	44.3	4.27	3.96	100	--	-76.4	-52.6
CHO	8.65	44.3	4.88				-83.5	
	Y2 GγS pK$_1$		Y2 GγS pK$_2$	Y2 GβS pK$_1$		Y2 GβS pK$_2$		
PAR	8.18	19.6	3.07	4.33	100	--	-56.0	-14.0
MBA	6.10	34.9	3.14	6.35	63.2	asymptote	-50.4	+57.4
POA	4.42	74.4	4.43	6.51	65.6	asymptote	-25.4	+33.3
HIPP	5.00	100	--	2.98	100	--	-22.5	-10.0
CHO	8.29	25.6	3.12	4.31	62.6	asymptote	-65.4	-36.0

pK= -log(K) where K is the estimated 50% decrease or increase of the specific binding relative to that without the guanine nucleotide. The GβS profiles with the Y2 receptor were best fit to the logistic, and other profiles to the biexponential model . The % K_1 is the percentage of specific binding in the higher affinity component. The % K_2 (lower affinity component or asymptote) is the difference (100 - % K$_1$). (Pro34)hPYY and hPYY(3-36) were used as the Y1 and Y2 tracer, respectively. The *GγS change* and *Gβ % change* show percent maximal change of total specific binding by the agent (inhibition (-) or activation (+)). For the Y2 MBA profile with GDP the pK$_1$ value was 6.16, representing +56.4% of the specific binding.

However, at least one-third of the Y1 binding is lost to blockade of Gα nucleotide sites by GβS.

Sensitivity to GγS of the Y2-selective binding (Figure 7B) was very diverse across the brain areas, with pKi values in nanomolar range for PAR and CHO-Y2 (albeit at less than 30% of total rundown), about 1 μM in MBA, and in 10-25 μM range in HIPP and POA. The Y2 binding in epithelial-type cells (CHO-Y2 binding shown in Figure 7B; very similar profiles were obtained for the rabbit kidney Y2 binding) has a significant component sensitive to low concentrations of GγS, which indicates involvement of Gi-type nucleotide sites. However, in most brain areas the Y2 binding is quite refractory to GγS, and could even be activated in response to the nucleotide (Figure 7B). The Y2 binding is strongly activated in particulates from hypothalamic areas by the irreversible nucleotide site antagonist GβS (Fig 7B). This however requires high concentrations of the nucleotide, which may point to association with Gq-type α subunits (e.g. (Hepler et al., 1993)). Occupancy of the nucleotide sites could activate the Y2 binding by detachment of area-specific inhibitors that also bind to the receptor. The Y2R may interact with different Gα subunits with different rates of activation (e.g. fast for the Gi, and slow, delayed and protracted for Gq, also depending on the activating proteins involved). Also there could be important differences in abundance of particular G-protein classes in discrete hypothalamic and other limbic areas. The paradoxical activation of Y2 binding by blockade of Gα nucleotide sites could also indicate that association with effector(s) supports agonist attachment to the receptor, e.g. by reducing lateral mobility of the receptor (Prenner et al., 2007; Rees et al., 1984).

2.6. Masking and Aggregation as Factors Affecting the Activity of Y1 and Y2 Receptors

After labeling in particulates, a substantial portion of agonist binding to the rabbit kidney and hypothalamic Y2 sites is not dispersed by digitonin/cholate, and sediments as aggregates. The aggregates are much reduced by further incubation for 15 min at 23 $^{\circ}$ with 10 μM GγS, and the dispersed labeled receptors sediment as Gi- (and likely also Gβγ-) linked monomers, at 90-100 kDa (Figure 8B, C; note that incubation at 37 $^{\circ}$ at $\geq$10 μM GγS would result in a large rundown detachment of the agonist; see (Estes et al., 2008)). This disaggregation and the large unmasking by GβS (Figure 7B) could point to an area-linked difference in Y2 receptor compartmentalization.

Receptor masking by interaction with substratum and ECM proteins is strong with monolayers of CHO-Y2 cells, but not with CHO-Y1, CHO-Y4 and HEK293-Y5 monolayers (Parker et al., 2012b). The Y2 masking is also observed with dispersed hypothalamic cells (Parker et al., 2002c). The non-cytolytic unmasking of the CHO-Y2R is achieved by gentle mechanical dispersion, trivalent arsenicals, EDTA, steroid detergent digitonin and phospholipid ether edelfosine (Parker et al., 2001b; Parker et al., 2002c; Parker et al., 2012b) . These treatments however largely do not have physiological parallels, and also do not address intracellular (and intercellular) masking. We therefore examined unmasking of Y receptors in particulates by proteolysis.

Trypsinization at 1-3 μg/ml did not produce substantial changes in K_{diss} at any of the employed concentrations of trypsin for any of the particulate samples (data not shown). The Y1 binding to piriform cortex particulates was somewhat decreased at the highest concentration of trypsin. The CHO-Y2 binding was only about 15% increased, in keeping with the bulk of unmasking being produced by cell dispersion. The Y2 unmasking was about 40% in hypothalamic particulates (Figure 8A) and also in piriform particulates (not shown). A very large Y2 unmasking was found with rabbit kidney particulates. The rabbit kidney or rat anterior hypothalamic area (AHA; the pooled POA and MBA areas) Y2 receptors also have a large fraction of sites in aggregates that are not solubilized by digitonin plus cholate (10 mM each). About 60% of these sites are disaggregated as 90-100 kDa material by 10 μM GγS within 15 min at 24 °C, and with kidney particulates this activation is accompanied by a large conversion of 180 kDa heteropentamers to monomers (Figure 8B). Trypsin at $\leq$ 300 ng/ml disperses the kidney aggregates to mainly release the heteropentamers (data not shown). The lack of receptor inactivation by trypsin at concentrations that degrade free NPY (inset of Figure 8A) is also a novel observation and should have methodological value.

Disaggregation by the phospholipase C inhibitor U73122 was small, but consistently present in both kidney and AHA assays (Figure 8B, C), and obviously merits further study.

2.7. Rundown and Autolysis as Factors Affecting the Apparent Levels of Y1 and Y2 Receptors

Evaluation of the expression of receptors in the brain, including the Y receptors, could encounter problems related to autolysis, especially with human tissue taken from cadavers (Caberlotto et al., 1997). The Y2R is both better protected from proteinase activity (probably as related to masking), and experiences a lower rundown (which may depend on several factors, including differences in transducing partners and effectors).

Particulates isolated after rat tissue incubation of 30 min at 37 ° showed a substantial changes in Y1 and Y2 binding (Figure 9). There were substantial ic_{50} changes for both the Y1 and the Y2 binding, and this was pronounced in AHA particulates (Fig. 9A).

The high affinity binding (B_0) was accordingly much reduced by autolysis (Figure 9B). The magnitude of changes indicates that any quantitation of NPY receptors (and especially the Y1) by agonist binding, which is strongly dependent on G-protein and effector coupling to the receptors, would be seriously compromised by preincubation of tissue at temperatures in physiological range. The Y1R binding is obviously more affected by autolysis than that of the Y2R.

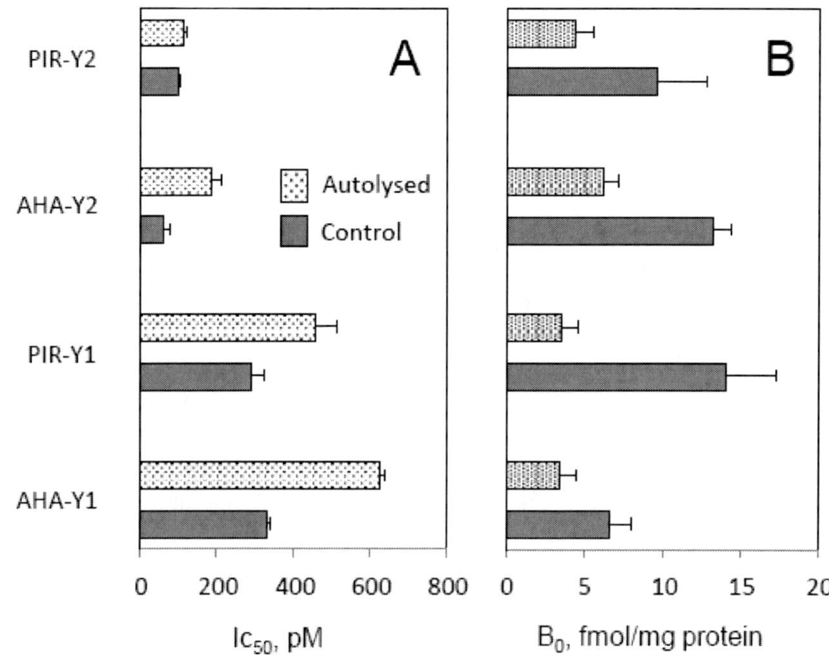

Figure 9. Rundown of rat brain Y1 and Y2 receptors due to tissue preincubation at 37^0. Shown are the ic_{50} and B_0 values for particulates from control tissue and from tissue incubated / autolysed for 30 min at 37 °C prior to particle isolation. (A) The ic_{50} values (n = 4) are from single-component fits of the competitions of 25 pM [^{125}I]-labeled selective agonists (Pro34)hPYY (Y1R) and hPYY(3-36) (Y2R) with six concentrations of the respective parent peptides in the range of 3-1000 pM. (B) The B_0 values (n = 4) are the estimates for the binding in absence of competitors.

This may indicate a generally faster transduction due to lower masking, and also a lower stability of Y1 agonist association with its two-prong binding site.

2.8. NPY Strongly Associates with Non-Receptor Targets and Can Block Surface Y1 Sites

In epithelial-type and neuroblastoma cell cultures, activities of Y receptors, and especially of the Y1 subtype, are known to be blocked by prolonged exposure to high concentrations of primary peptidic agonists ((Parker et al., 2007a; Sah et al., 2005); partly reproduced in Fig 10). This could be connected to the large non-specific binding of NPY ((Parker et al., 2012b); partly reproduced in Figure 11).

As we have shown previously (Parker et al., 2007a; Sah et al., 2005), NPY at $\geq$3 nM apparently blocks the surface CHO-Y1 and SK-N-MC Y1 sites, with ic_{50} values similar to peptidic antagonists (Figure 10A), and also strongly inhibits the Y1R internalization (Figure 10B). This is shared by the Y1-selective agonist (Leu31,Pro34)hNPY. Peptide YY and the PYY-derived Y1 agonist (Leu31,Pro34)hPYY even at 300 nM produced less than 50% Y1 blockade (Figure 10A), and apparently somewhat stimulated internalization of the Y1R (Figure 10B). With the CHO-Y2 receptor, however, pPYY and hNPY both produced less than 50% blockade or decrease in Y2R internalization at 300 nM (Figure 10C, D). The Y2-selective agonist hPYY(3-36) was significantly less blocking; however this peptide is known to block the particulate Y2 receptors at high inputs (Dautzenberg and Neysari, 2005). The above findings could reflect interaction of NPY with membrane-associated receptor partners that can be tolerated by the Y2R, but not by the two-prong Y1R binding site. The above reasoning could be supported by profiles of NPY and PYY non-specific binding to CHO-Y1 cell particulates (Figure 11, adapted from Figure 4 in (Parker et al., 2012b)) at 50 pM of $[^{125}I]$-labeled and 100 nM of the unlabeled respective peptides. With these inputs, the non-specific binding of NPY per mg particle protein is about 2.5 times larger than that of PYY, and upon solubilization sediments chiefly between 75 and 160 kDa. The much smaller non-specific binding of PYY quite lacks the range past about 120 kDa. Partners of NPY obviously can include a variety of large membrane-dwelling molecules, including the α subunits of voltage-gated channels, and membrane-associating enzymes (such as phospholipases C), and there could be interactions with large GTPases, e.g. dynamins (Fan et al., 2001). Many of the above molecules have molecular weights in the 100-160 kDa range. Notably, the non-specific binding of NPY is quite sensitive to monovalent cations (Parker et al., 2011), which should reflect association with basic residues in protein sequences.

3. General Discussion

In a broad sense, activities of peptidic agonists, and especially of synaptically released neuropeptides, could significantly depend on the constitutive element of ionic segregation, and on the local concentration after discharge.

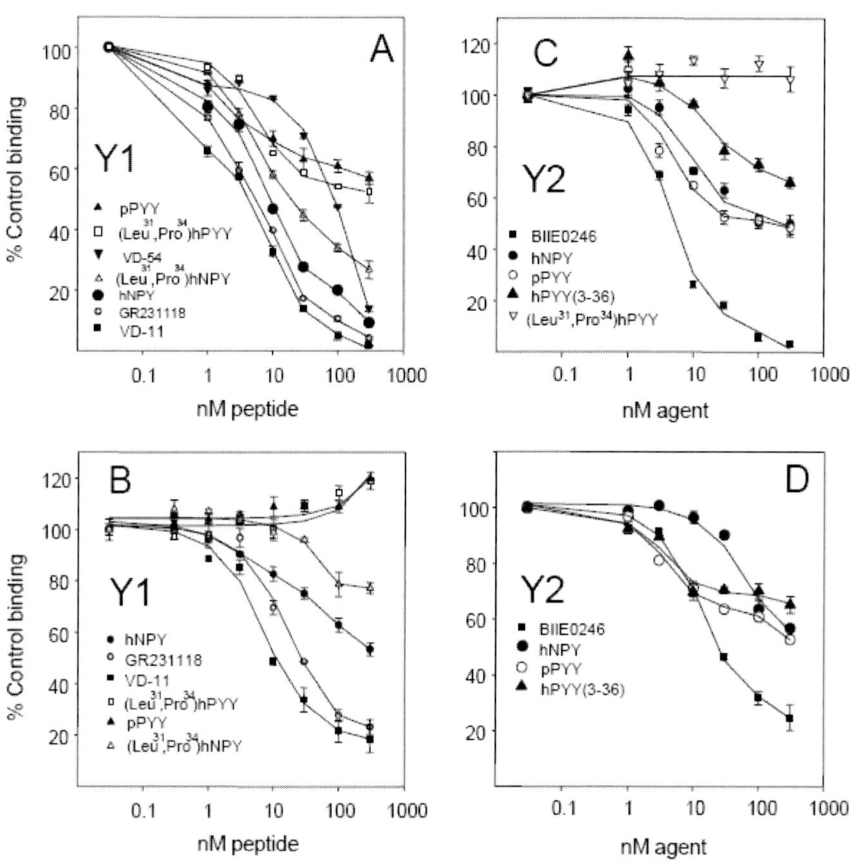

Figure10. NPY and NPY-derived Y1-selective agonist at high concentrations block the Y1 surface sites and internalization in CHO cells. (Partly adapted from (Parker et al., 2007a)). (A) Cell surface binding of [^{125}I]hPYY(1-36) (20 min at 37 $^{\circ}$) to CHO-Y1 cells after cell exposure to the shown concentrations of agents for 45 min at 37 $^{\circ}$, and a recovery incubation without the agents for 30 min at 37 $^{\circ}$. (B) The corresponding internalization of [^{125}I]hPYY(1-36). (C) Cell surface binding of [^{125}I]hPYY(3-36) to CHO-hY2 cells, with treatment conditions described in (A). (D) The corresponding internalization of [^{125}I]hPYY(3-36).   The blockade is not observed with PYY or a PYY-derived Y1 agonist. The blockade is also not found with CHO-Y2 receptor (which is however completely blocked by the irreversible Y2 antagonist BIIE0246). The Y1 blockade by NPY is similar to those by peptidic Y1 antagonists GR231118 and VD-11. The Y1-selective non-peptide antagonist BIBP3226 was blocking with an ic$_{50}$ of about 3 μM (not shown). The Y1 effects were reproduced with SK-N-MC cells.

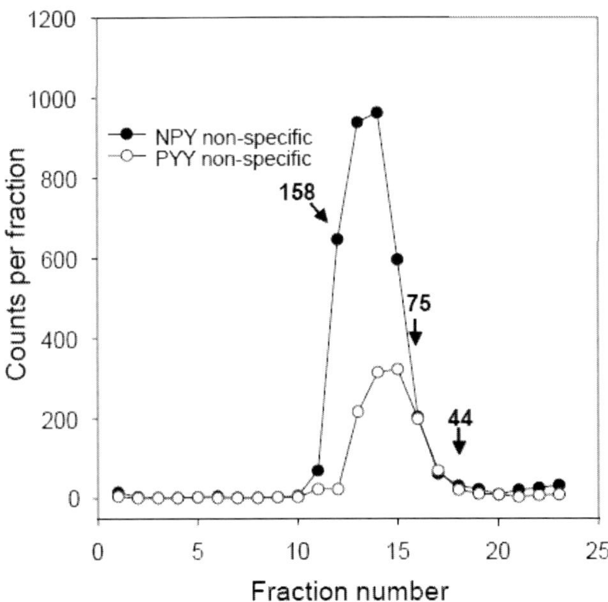

Figure 11. A comparison of the non-specific binding of hNPY and pPYY to CHO cells expressing the human Y1 receptor (adapted from Fig. 4 in {Parker, 2011 #478}). The ^{125}I –labeled agonists were added to particulates at 50 pM with 100 nM of the corresponding unlabeled parent peptide. After labeling for 30 min at 25 °C, the particulates were sedimented, surface-washed, and lysed with 10 mM each digitonin and cholate, followed by sedimentation through 10-30% sucrose gradients for 24 h at 218,000 x g_{max}, and gradient fractionation. The non-specific NPY binding was about 12% of total binding. The numbers indicate the sedimentation of the molecular weight standards, in kDa ((Parker et al., 2011).

The former is highly implemented throughout the natural ligand-type peptides (Parker et al., 2012a), and the latter should relate more specifically to abundant neuropeptides, including NPY (Allen et al., 1983) and opioid peptides (Kosanam et al., 2009). A discharge of NPY may provide sufficient inputs to extracellular domains of clustered cation channels (Giesbrecht et al., 2010). This effect could also be supported and maintained by more specific and longer-lasting activation of transducers and effectors coupled to NPY receptors and also interacting with intracellular domains of channel proteins.

The apparent antagonism of Y1 and Y2 could be based on [1] compartmentalization / slow activation of Y2 sites (Figure 7B), [2] different activation of transducers (e.g., Gq activation low by the Y1R (Parker et al., 2008b), and strong by the Y2R (Grouzmann et al., 2001)), [3] the overall balance of Y1and Y2 sites. Areas with sufficient Y1 densities (which include most forebrain areas) and with low (grey matter) or strongly compart-mentalized (hypothalamus) Y2 receptors would quickly respond to NPY by lowering cAMP production and glucose production, creating hunger stimuli. Dorsomedial vagal nuclei in the brainstem may have an enrichment in Y2 sites (which however is not clear from the current evidence) and could produce amplified signals in tune with the rise of PYY(3-36) during feeding. The inhibition of feeding by Y2 agonists seems to be confined to systemically supplied 3-36 peptides (Kanatani et al., 2000; Batterham et al., 2002; Martin et al., 2004; Vrang et al., 2006; Balasubramaniam et al., 2007; Moriya et al., 2009; Rodgers et al., 2010). Intracerebrally administered 3-36 peptides in most studies are found to stimulate feeding in the rat. NPY(3-36) has a much larger overall interaction potency than PYY(3-36) (Parker et al., 2011) and could be significantly more orexigenic. Intracerebral NPY(3-36) is consistently reported to stimulate feeding in the rat at a potency similar to that of NPY(1-36) (Wieland et al., 1998; Haynes et al., 1998; Flynn et al., 1999; Aldegunde and Mancebo, 2006; Travers et al., 2010). NPY(3-36), unlike PYY(3-36), is not normally present at appreciable levels in blood, and systemic administration of NPY(3-36) hence was not studied. Intraventricular application of PYY(3-36) could stimulate feeding in the rat (Kanatani et al., 2000; Corp et al., 2001a), although it could be inhibitory in rhesus monkey (Papadimitriou et al., 2007). Given the similar potency of NPY(3-36) and PYY(3-36) in the Y2 receptor binding and activation of Gi α subunits (Parker and Balasubramaniam, 2008), one would expect a stimulatory role for the endogenously generated NPY(3-36), which is one of the major products of NPY cleavage in rat brain (Frerker et al., 2007; Grandt et al., 1996) and likely to persist in csf over long periods (figures 2-3).

The Y2 and Y4 receptors in the dorsal motor nucleus of the vagus and in area postrema (Dumont et al., 2007) of the brainstem raphe nuclei should be important in feeding inhibition in the rat (Chandarana and Batterham, 2008; Dockray and Burdyga, 2011). These areas are clearly accessible to systemic peptides. PYY(3-36), which by action of dipeptidylpeptidase IV increases to 50% of all PYY forms during feeding (Batterham et al., 2002), also has very high Y2 binding activity, and may subtractively overpower the vagal Y1 receptor. The vagal Y1 and Y2 receptors are not likely to receive high inputs of NPY. There also could be regulation of the Y2R by cholecystokinin in the dorsal vagal projections (Burdyga et al., 2008), which may not be present in the forebrain.

The above consideration is based entirely on the classical activation of Y1 and Y2 receptors by NPY, PYY and their 3-36 peptides, to initiate signal transduction e.g. via Gi α subunits. However, as shown in Table 2, both Y agonists and their receptors possess homoionic domains that may directly interact with well-identified domains and sectors of cation channels, e.g. in conditions of synaptic discharge. This can help explain the large variation in responses of ion channels to Y agonists (see section 1.9). NPY receptors could have broad scope of association with channels and transporters. Association through extracellular domains could be significant especially in the case of the Y2 receptor (Parker et al., 2008c). The Y5 receptor is obviously likely to interact extensively through large basic zippers in the third intracellular loop, but this was not researched thus far. The Y1 and Y2 receptors both may interact with ion channels. This can use strongly homoionic (and especially homobasic) intracellular tracts of the two receptors (Table 2). However, actions of NPY that trigger the ion channels could also be direct, without involvement of Y receptors. NPY is one of the most abundant brain hormones (Adrian et al., 1983; Allen et al., 1983), and its interaction with any partners could be promoted simply by mass-action. The non-specific binding of NPY is quite sensitive to monovalent cations (Parker et al., 2011) indicating association with basic residues in 80-160 kDa proteins (Figure 11), which could include channels, transporters and enzymes.

NPY importation via internalization of the Y1 receptor may make NPY directly available to G-protein α subunits (or other GTPases) after separation of the internalized receptor from a Gi α subunit. The affinity of NPY for the receptor should strongly decrease upon separation of the receptor from G-protein, and the released NPY could then attach to the transducer. Most of the non-specific binding of NPY sediments at 80-160 kDa upon solubilization; the non-specific binding of PYY is about 2.5 times lower, and sediments at 75-120 kDa, indicating a difference in partners (Figure 11). Partners of NPY could also include effectors such as channels, transporters, protein kinases and phosphatases, as well as other receptors. This should be examined in connection to inhibition of neurotransmitter secretion, e.g. of glutamate release by NPY(Wang, 2005).

Both stimulatory and inhibitory activities of NPY upon channels via the Y1R and the Y2R have been reported in a variety of cells, tissues and animal models. This might depend on cell and tissue type as well as on levels of expression of the Y receptors and of the respective channels, which in most studies was not specifically addressed. These interactions could involve the intracellular domains of both Y receptors and channels (voltage-gated or ligand-gated) and transporters, probably relying on homoionic matching. This could be implemented at the level of ER/Golgi completion or of the endosomal recycling. Some candidates for this type of matching are listed in Table 2. Ability of the Y2 receptor to inhibit both orexigenic and anorexigenic arcuate neurons (Acuna-Goycolea and van den Pol, 2005) adds weight to this possibility. The similarity of effects of NPFF2 and Y2 agonists upon the N-type Ca2+ currents (Mollereau et al., 2007) can have the same roots.

Homoionic segments in extracellular domains of voltage-gated and ligand-gated channels may serve at least for transient docking of NPY. The voltage-sensing S4 helices (which are readily accessible to cell surface interactants (Tombola et al., 2005)) and the S6-preceding extracellular domains of voltage-gated channels have strongly homoionic constitution (Table 2). Analogous with conotoxins (French et al., 1996) and agatoxins (Weiss and Burgoyne, 2001), NPY can interact with these tracts in calcium channels (Qian et al., 1997) in conditions of synaptic discharge, targeting the opposite-charged homoionic zippers. The large avidity of NPY for phospholipids (McLean et al., 1990) could lead to significant accumulation in the bilayer (see also (Bader et al., 2002; Lerch et al., 2004)), helping contacts with any transmembrane and in-membrane tracts. The specific NPY binding depends significantly on residues in the transmembrane helices of the Y1 receptor (Du et al., 1997; Sautel et al., 1996) and the Y2 receptor (Fallmar et al., 2011)). The strongly homobasic voltage sensors (Table 2) could be targets for both extracellular and in-membrane NPY. Also, there could be quasi-irreversible association with membrane-ensconced or membrane-associating proteins, as evidenced by the large and stable non-specific binding of NPY to 80-160 kDa targets (Figure 11; see also Figure 4 in (Parker et al., 2011)).

The large homoacidic tracts of cytosolic effectors (Table 3) should react mainly with homobasic sections of the intracellular domains of the receptors (shown in Table 2 for the Y receptors). The Y1, Y2 and Y4 receptors all transduce to phospholipase C effectors (Grouzmann et al., 2001; Parker et al., 1999; Parker et al., 1998b; Selbie et al., 1997). Here it also should be mentioned that the large homoacidic N-terminal extracellular tracts of many GPCRs, and especially of the chemokine receptors, are obvious candidates for direct interactions with Y peptides. This is currently being studied in our laboratories.

Evidence presented in this chapter clearly indicates the need for extensive examination of both direct and receptor-linked activities of NPY (and Y peptides in general) in terms of interactions involving especially the homoionic aspects of transducers and effectors, including channels and transporters. This should be supported by a detailed characterization of the participating Y receptors, especially in brainstem / medullar and gastrointestinal locations.

References

Aakerlund L, Gether U, Fuhlendorff J, Schwartz TW and Thastrup O (1990) Y1 receptors for neuropeptide Y are coupled to mobilization of intracellular calcium and inhibition of adenylate cyclase. *FEBS Lett.* 260:73-78.

Abid K, Rochat B, Lassahn PG, Stocklin R, Michalet S, Brakch N, Aubert JF, Vatansever B, Tella P, De Meester I and Grouzmann E (2009) Kinetic study of neuropeptide Y (NPY) proteolysis in blood and identification of NPY3-35: a new peptide generated by plasma kallikrein. *J. Biol. Chem.* 284:24715-24724.

Acuna-Goycolea C and van den Pol AN (2005) Peptide YY(3-36) inhibits both anorexigenic proopiomelanocortin and orexigenic neuropeptide Y neurons: implications for hypothalamic regulation of energy homeostasis. *J. Neurosci.* 25:10510-10519.

Adrian TE, Allen JM, Bloom SR, Ghatei MA, Rossor MN, Roberts GW, Crow TJ, Tatemoto K and Polak JM (1983) Neuropeptide Y distribution in human brain. *Nature* 306:584-586.

Aicher SA, Springston M, Berger SB, Reis DJ and Wahlestedt C (1991) Receptor-selective analogs demonstrate NPY/PYY receptor heterogeneity in rat brain. *Neurosci. Lett.* 130:32-36.

Aldegunde M and Mancebo M (2006) Effects of neuropeptide Y on food intake and brain biogenic amines in the rainbow trout (Oncorhynchus mykiss). *Peptides* 27:719-727.

Allen YS, Adrian TE, Allen JM, Tatemoto K, Crow TJ, Bloom SR and Polak JM (1983) Neuropeptide Y distribution in the rat brain. *Science* 221:877-879.

Arcemisbehere L, Sen T, Boudier L, Balestre MN, Gaibelet G, Detouillon E, Orcel H, Mendre C, Rahmeh R, Granier S, Vives C, Fieschi F, Damian M, Durroux T, Baneres JL and Mouillac B (2011) Leukotriene BLT2 receptor monomers activate the G(i2) GTP-binding protein more efficiently than dimers. *J. Biol. Chem.* 285:6337-6347.

Arumugam R, Fleenor D and Freemark M (2007) Lactogenic and somatogenic hormones regulate the expression of neuropeptide Y and cocaine- and amphetamine-regulated transcript in rat insulinoma (INS-1) cells:

interactions with glucose and glucocorticoids. *Endocrinology* 148:258-267.

Bader R, Rytz G, Lerch M, Beck-Sickinger AG and Zerbe O (2002) Key motif to gain selectivity at the neuropeptide Y5-receptor: structure and dynamics of micelle-bound [Ala31, Pro32]-NPY. *Biochemistry* 41:8031-8042.

Balasubramaniam A, Joshi R, Su C, Friend LA and James JH (2007) Neuropeptide Y (NPY) Y2 receptor-selective agonist inhibits food intake and promotes fat metabolism in mice: combined anorectic effects of Y2 and Y4 receptor-selective agonists. *Peptides* 28:235-240.

Baneres JL and Parello J (2003) Structure-based analysis of GPCR function: evidence for a novel pentameric assembly between the dimeric leukotriene B4 receptor BLT1 and the G-protein. *J. Mol .Biol.* 329:815-829.

Batterham RL, Cowley MA, Small CJ, Herzog H, Cohen MA, Dakin CL, Wren AM, Brynes AE, Low MJ, Ghatei MA, Cone RD and Bloom SR (2002) Gut hormone PYY(3-36) physiologically inhibits food intake. *Nature* 418:650-654.

Beck-Sickinger AG, Wieland HA, Wittneben H, Willim KD, Rudolf K and Jung G (1994) Complete L-alanine scan of neuropeptide Y reveals ligands binding to Y1 and Y2 receptors with distinguished conformations. *Eur. J. Biochem.* 225:947-958.

Berglund MM, Fredriksson R, Salaneck E and Larhammar D (2002) Reciprocal mutations of neuropeptide Y receptor Y2 in human and chicken identify amino acids important for antagonist binding. *FEBS Lett.* 518:5-9.

Berglund MM, Schober DA, Esterman MA and Gehlert DR (2003) Neuropeptide Y Y4 receptor homodimers dissociate upon agonist stimulation. *J. Pharmacol. Exp. Ther.* 307:1120-1126.

Berstein G, Blank JL, Jhon DY, Exton JH, Rhee SG and Ross EM (1992) Phospholipase C-beta 1 is a GTPase-activating protein for Gq/11, its physiologic regulator. *Cell* 70:411-418.

Bettio A, Gutewort V, Poppl A, Dinger MC, Zschornig O, Klaus A, Toniolo C and Beck-Sickinger AG (2002) Electron paramagnetic resonance backbone dynamics studies on spin-labelled neuropeptide Y analogues. *J. Pept. Sci.* 8:671-682.

Biddlecome GH, Berstein G and Ross EM (1996) Regulation of phospholipase C-beta1 by Gq and m1 muscarinic cholinergic receptor. Steady-state balance of receptor-mediated activation and GTPase-activating protein-promoted deactivation. *J. Biol. Chem.* 271:7999-8007.

Blundell TL, Pitts JE, Tickle IJ, Wood SP and Wu CW (1981) X-ray analysis (1. 4-A resolution) of avian pancreatic polypeptide: Small globular protein hormone. *Proc. Natl. Acad. Sci. U S A* 78:4175-4179.

Brown NA, McAllister G, Weinberg D, Milligan G and Seabrook GR (1995) Involvement of G-protein alpha il subunits in activation of G-protein gated inward rectifying K+ channels (GIRK1) by human NPY1 receptors. *Br. J. Pharmacol.* 116:2346-2348.

Burdakov D, Luckman SM and Verkhratsky A (2005) Glucose-sensing neurons of the hypothalamus. *Philos. Trans. R. Soc. Lond. B. Biol. Sci.* 360:2227-2235.

Burdyga G, de Lartigue G, Raybould HE, Morris R, Dimaline R, Varro A, Thompson DG and Dockray GJ (2008) Cholecystokinin regulates expression of Y2 receptors in vagal afferent neurons serving the stomach. *J. Neurosci.* 28:11583-11592.

Caberlotto L, Fuxe K and Hurd YL (2000) Characterization of NPY mRNA-expressing cells in the human brain: co-localization with Y2 but not Y1 mRNA in the cerebral cortex, hippocampus, amygdala, and striatum. *J. Chem. Neuroanat.* 20:327-337.

Caberlotto L, Fuxe K, Sedvall G and Hurd YL (1997) Localization of neuropeptide Y Y1 mRNA in the human brain: abundant expression in cerebral cortex and striatum. *Eur. J. Neurosci.* 9:1212-1225.

Camina JP, Carreira MC, El Messari S, Llorens-Cortes C, Smith RG and Casanueva FF (2004) Desensitization and endocytosis mechanisms of ghrelin-activated growth hormone secretagogue receptor 1a. *Endocrinology* 145:930-940.

Campbell RE, Smith MS, Allen SE, Grayson BE, Ffrench-Mullen JM and Grove KL (2003) Orexin neurons express a functional pancreatic polypeptide Y4 receptor. *J. Neurosci.* 23:1487-1497.

Castro A, Manso MJ and Anadon R (2003) Distribution of neuropeptide Y immunoreactivity in the central and peripheral nervous systems of amphioxus (Branchiostoma lanceolatum Pallas). *J. Comp. Neurol.* 461:350-361.

Chandarana K and Batterham R (2008) Peptide YY. *Curr. Opin. Endocrinol. Diabetes Obes.* 15:65-72.

Chatenet D, Cescato R, Waser B, Erchegyi J, Rivier JE and Reubi JC (2011) Novel dimeric DOTA-coupled peptidic Y1-receptor antagonists for targeting of neuropeptide Y receptor-expressing cancers. *EJNMMI Res.* 1:21.

Christie AE, Chapline MC, Jackson JM, Dowda JK, Hartline N, Malecha SR and Lenz PH (2011) Identification, tissue distribution and orexigenic activity of neuropeptide F (NPF) in penaeid shrimp. *J. Exp. Biol.* 214:1386-1396.

Cline MA, Newmyer BA and Smith ML (2009) The anorectic effect of neuropeptide AF is associated with satiety-related hypothalamic nuclei. *J. Neuroendocrinol.* 21:595-601.

Connor M, Yeo A and Henderson G (1997) Neuropeptide Y Y2 receptor and somatostatin sst2 receptor coupling to mobilization of intracellular calcium in SH-SY5Y human neuroblastoma cells. *Br. J. Pharmacol.* 120:455-463.

Corp ES, Greco B, Powers JB, Marin Bivens CL and Wade GN (2001a) Neuropeptide Y inhibits estrous behavior and stimulates feeding via separate receptors in Syrian hamsters. *Am. J. Physiol. Regul. Integr. Comp. Physiol.* 280:R1061-1068.

Corp ES, McQuade J, Krasnicki S and Conze DB (2001b) Feeding after fourth ventricular administration of neuropeptide Y receptor agonists in rats. *Peptides* 22:493-499.

Crespi EJ, Vaudry H and Denver RJ (2004) Roles of corticotropin-releasing factor, neuropeptide Y and corticosterone in the regulation of food intake in Xenopus laevis. *J. Neuroendocrinol.* 16:279-288.

Crowley WR, Shah GV, Carroll BL, Kennedy D, Dockter ME and Kalra SP (1990) Neuropeptide-Y enhances luteinizing hormone (LH)-releasing hormone-induced LH release and elevations in cytosolic Ca2+ in rat anterior pituitary cells: evidence for involvement of extracellular Ca2+ influx through voltage-sensitive channels. *Endocrinology* 127:1487-1494.

D'Angelo I and Brecha NC (2004) Y2 receptor expression and inhibition of voltage-dependent Ca2+ influx into rod bipolar cell terminals. *Neuroscience* 125:1039-1049.

D'Ursi AM, Albrizio S, Di Fenza A, Crescenzi O, Carotenuto A, Picone D, Novellino E and Rovero P (2002) Structural studies on Hgr3 orphan receptor ligand prolactin-releasing peptide. *J Med Chem* 45:5483-5491.

Daniels AJ, Matthews JE, Slepetis RJ, Jansen M, Viveros OH, Tadepalli A, Harrington W, Heyer D, Landavazo A, Leban JJ and Spaltenstein A (1995) High-affinity neuropeptide Y receptor antagonists. *Proc. Natl. Acad. Sci. U S A* 92:9067-9071.

Daniels AJ, Matthews JE, Viveros OH, Leban JJ, Cory M and Heyer D (1995) Structure-activity relationship of novel pentapeptide neuropeptide Y receptor antagonists is consistent with a noncontinuous epitope for ligand-receptor binding. *Mol. Pharmacol.* 48:425-432.

Dautzenberg FM and Neysari S (2005) Irreversible binding kinetics of neuropeptide Y ligands to Y2 but not to Y1 and Y5 receptors. *Pharmacology* 75:21-29.

Dinger MC, Bader JE, Kobor AD, Kretzschmar AK and Beck-Sickinger AG (2003) Homodimerization of neuropeptide y receptors investigated by fluorescence resonance energy transfer in living cells. *J. Biol. Chem.* 278:10562-10571.

Dockray GJ and Burdyga G (2011) Plasticity in vagal afferent neurones during feeding and fasting: mechanisms and significance. *Acta Physiol. (Oxf)* 201:313-321.

Dougherty DA (1996) Cation-p interactions in chemistry and biology: a new view of benzene, Phe, Tyr, and Trp. *Science* 271:163-168.

Du P, Salon JA, Tamm JA, Hou C, Cui W, Walker MW, Adham N, Dhanoa DS, Islam I, Vaysse PJ, Dowling B, Shifman Y, Boyle N, Rueger H, Schmidlin T, Yamaguchi Y, Branchek TA, Weinshank RL and Gluchowski C (1997) Modeling the G-protein-coupled neuropeptide Y Y1 receptor agonist and antagonist binding sites. *Protein Eng.* 10:109-117.

Dumont Y, Fournier A and Quirion R (1998) Expression and characterization of the neuropeptide Y Y5 receptor subtype in the rat brain. *J. Neurosci.* 18:5565-5574.

Dumont Y, Fournier A, St-Pierre S and Quirion R (1993) Comparative characterization and autoradiographic distribution of neuropeptide Y receptor subtypes in the rat brain. *Journal of Neuroscience* 13:73-86.

Dumont Y, Fournier A, St-Pierre S and Quirion R (1995) Characterization of neuropeptide y binding sites in rat brain membrane preparations using [I-125][Leu(31),Pro(34)]peptide YY and [I-125]peptide YY3-36 as selective Y-1 and Y-2 radioligands. *J. Pharmacol. Exp. Therap.* 272:673-680.

Dumont Y, Moyse E, Fournier A and Quirion R (2007) Distribution of peripherally injected peptide YY ([125I] PYY (3-36)) and pancreatic polypeptide ([125I] hPP) in the CNS: enrichment in the area postrema. *J. Mol. Neurosci.* 33:294-304.

Estes AM, Wong YY, Parker MS, Sallee FR, Balasubramaniam A and Parker SL (2008) Neuropeptide Y (NPY) Y2 receptors of rabbit kidney cortex are largely dimeric. *Regul. Pept.* 150:88-94.

Fallmar H, Kerberg H, Gutierrez-de-Teran H, Lundell I, Mohell N and Larhammar D (2011) Identification of positions in the human neuropeptide Y/peptide YY receptor Y2 that contribute to pharmacological differences between receptor subtypes. *Neuropeptides* 45:293-300.

Fan GH, Yang W, Sai J and Richmond A (2001) Phosphorylation-independent association of CXCR2 with the protein phosphatase 2A core enzyme. *J. Biol. Chem.* 276:16960-16968.

Filizola M, Wang SX and Weinstein H (2006) Dynamic models of G-protein coupled receptor dimers: indications of asymmetry in the rhodopsin dimer from molecular dynamics simulations in a POPC bilayer. *J. Comput. Aided Mol. Des.* 20:405-416.

Flynn MC, Plata-Salaman CR and Ffrench-Mullen JM (1999) Neuropeptide Y-related compounds and feeding. *Physiol. Behav.* 65:901-905.

Ford CE, Skiba NP, Bae H, Daaka Y, Reuveny E, Shekter LR, Rosal R, Weng G, Yang CS, Iyengar R, Miller RJ, Jan LY, Lefkowitz RJ and Hamm HE (1998) Molecular basis for interactions of G protein betagamma subunits with effectors. *Science* 280:1271-1274.

Freitag C, Svendsen AB, Feldthus N, Lossl K and Sheikh SP (1995) Coupling of the human Y2 receptor for neuropeptide Y and peptide YY to guanine nucleotide inhibitory proteins in permeabilized SMS-KAN cells. *J. Neurochem.* 64:643-650.

French RJ, Prusak-Sochaczewski E, Zamponi GW, Becker S, Kularatna AS and Horn R (1996) Interactions between a pore-blocking peptide and the voltage sensor of the sodium channel: an electrostatic approach to channel geometry. *Neuron* 16:407-413.

Frerker N, Wagner L, Wolf R, Heiser U, Hoffmann T, Rahfeld JU, Schade J, Karl T, Naim HY, Alfalah M, Demuth HU and von Horsten S (2007) Neuropeptide Y (NPY) cleaving enzymes: structural and functional homologues of dipeptidyl peptidase 4. *Peptides* 28:257-268.

Gao J, Ghibaudi L and Hwa JJ (2004) Selective activation of central NPY Y1 vs. Y5 receptor elicits hyperinsulinemia via distinct mechanisms. *Am. J. Physiol. Endocrinol. Metab.* 287:E706-711.

Garland AM, Grady EF, Payan DG, Vigna SR and Bunnett NW (1994) Agonist-induced internalization of the substance P (NK1) receptor expressed in epithelial cells. *Biochem. J.* 303 (Pt 1):177-186.

Gehlert DR, Schober DA, Beavers L, Gadski R, Hoffman JA, Smiley DL, Chance RE, Lundell I and Larhammar D (1996) Characterization of the peptide binding requirements for the cloned human pancreatic polypeptide-preferring receptor. *Mol. Pharmacol.* 50:112-118.

Gehlert DR, Schober DA, Gackenheimer SL, Beavers L, Gadski R, Lundell I and Larhammar D (1997) [125I]Leu31, Pro34-PYY is a high affinity radioligand for rat PP1/Y4 and Y1 receptors: evidence for heterogeneity in pancreatic polypeptide receptors. *Peptides* 18:397-401.

Gehlert DR, Schober DA, Morin M and Berglund MM (2007) Co-expression of neuropeptide Y Y1 and Y5 receptors results in heterodimerization and altered functional properties. *Biochem. Pharmacol.* 74:1652-1664.

Gierschik P, Milligan G, Pines M, Goldsmith P, Codina J, Klee W and Spiegel A (1986) Use of specific antibodies to quantitate the guanine nucleotide-binding protein Go in brain. *Proc. Natl. Acad. Sci. U S A* 83:2258-2262.

Giesbrecht CJ, Mackay JP, Silveira HB, Urban JH and Colmers WF (2010) Countervailing modulation of Ih by neuropeptide Y and corticotrophin-releasing factor in basolateral amygdala as a possible mechanism for their effects on stress-related behaviors. *J. Neurosci.* 30:16970-16982.

Grady EF, Slice LW, Brant WO, Walsh JH, Payan DG and Bunnett NW (1995) Direct observation of endocytosis of gastrin releasing peptide and its receptor. *J. Biol. Chem.* 270:4603-4611.

Grandt D, Schimiczek M, Rascher W, Feth F, Shively J, Lee TD, Davis MT, Reeve JR, Jr. and Michel MC (1996) Neuropeptide Y 3-36 is an endogenous ligand selective for Y2 receptors. *Regul. Pept.* 67:33-37.

Grandt D, Teyssen S, Schimiczek M, Reeve JR, Jr., Feth F, Rascher W, Hirche H, Singer MV, Layer P, Goebell H and et al. (1992) Novel generation of hormone receptor specificity by amino terminal processing of peptide YY. *Biochem. Biophys. Res. Commun.* 186:1299-1306.

Grouzmann E, Meyer C, Burki E and Brunner H (2001) Neuropeptide Y Y2 receptor signalling mechanisms in the human glioblastoma cell line LN319. *Peptides* 22:379-386.

Grunewald S, Reilander H and Michel H (1996) In vivo reconstitution of dopamine D2S receptor-mediated G protein activation in baculovirus-infected insect cells: preferred coupling to Gi1 versus Gi2. *Biochemistry* 35:15162-15173.

Hasstedt SJ, Xin Y, Hopkins PN and Hunt SC Two-dimensional, sex-specific autosomal linkage scan of the number of sodium pump sites. *J. Hypertens* 28:740-747.

Haynes AC, Arch JR, Wilson S, McClue S and Buckingham RE (1998) Characterisation of the neuropeptide Y receptor that mediates feeding in the rat: a role for the Y5 receptor? *Regul. Pept.* 75-76:355-361.

Hepler JR, Kozasa T, Smrcka AV, Simon MI, Rhee SG, Sternweis PC and Gilman AG (1993) Purification from Sf9 cells and characterization of recombinant Gq alpha and G11 alpha. Activation of purified phospholipase C isozymes by G alpha subunits. *J. Biol. Chem.* 268:14367-14375.

Hernandez EJ, Whitcomb DC, Vigna SR and Taylor IL (1994) Saturable binding of circulating peptide YY in the dorsal vagal complex of rats. *Am. J. Physiol.* 266:G511-516.

Herzog H, Baumgartner M, Vivero C, Selbie LA, Auer B and Shine J (1993) Genomic organization, localization, and allelic differences in the gene for the human neuropeptide Y Y1 receptor. *J. Biol. Chem.* 268:6703-6707.

Herzog H, Darby K, Ball H, Hort Y, Beck-Sickinger A and Shine J (1997) Overlapping gene structure of the human neuropeptide Y receptor subtypes Y1 and Y5 suggests coordinate transcriptional regulation. *Genomics* 41:315-319.

Hilairet S, Belanger C, Bertrand J, Laperriere A, Foord SM and Bouvier M (2001) Agonist-promoted internalization of a ternary complex between calcitonin receptor-like receptor, receptor activity-modifying protein 1 (RAMP1), and beta-arrestin. *J. Biol. Chem.* 276:42182-42190.

Hinuma S, Habata Y, Fujii R, Kawamata Y, Hosoya M, Fukusumi S, Kitada C, Masuo Y, Asano T, Matsumoto H, Sekiguchi M, Kurokawa T, Nishimura O, Onda H and Fujino M (1998) A prolactin-releasing peptide in the brain. *Nature* 393:272-276.

Holmberg SK, Mikko S, Boswell T, Zoorob R and Larhammar D (2002) Pharmacological characterization of cloned chicken neuropeptide Y receptors Y1 and Y5. *J. Neurochem.* 81:462-471.

Hunt SC, Hasstedt SJ, Xin Y, Dalley BK, Milash BA, Yakobson E, Gress RE, Davidson LE and Adams TD (2011) Polymorphisms in the NPY2R gene show significant associations with BMI that are additive to FTO, MC4R, and NPFFR2 gene effects. *Obesity (Silver Spring)* 19:2241-2247.

Hurley-Gius KM and Neafsey EJ (1986) The medial frontal cortex and gastric motility: microstimulation results and their possible significance for the overall pattern of organization of rat frontal and parietal cortex. *Brain Res.* 365:241-248.

Jacques D, Dumont Y, Fournier A and Quirion R (1997) Characterization of neuropeptide Y receptor subtypes in the normal human brain, including the hypothalamus. *Neuroscience* 79:129-148.

Jhanwar-Uniyal M, Beck B, Jhanwar YS, Burlet C and Leibowitz SF (1993) Neuropeptide Y projection from arcuate nucleus to parvocellular division of paraventricular nucleus: specific relation to the ingestion of carbohydrate. *Brain Res.* 631:97-106.

Kanatani A, Mashiko S, Murai N, Sugimoto N, Ito J, Fukuroda T, Fukami T, Morin N, MacNeil DJ, Van der Ploeg LH, Saga Y, Nishimura S and Ihara M (2000) Role of the Y1 receptor in the regulation of neuropeptide Y-mediated feeding: comparison of wild-type, Y1 receptor-deficient, and Y5 receptor-deficient mice. *Endocrinology* 141:1011-1016.

Kane JK, Tanaka H, Parker SL, Yanagisawa M and Li MD (2000) Sensitivity of orexin-A binding to phospholipase C inhibitors, neuropeptide Y, and secretin. *Biochem. Biophys. Res. Commun.* 272:959-965.

Kastin AJ and Akerstrom V (1999) Nonsaturable entry of neuropeptide Y into brain. *Am. J. Physiol.* 276:E479-482.

Kim SK and Jacobson KA (2006) Computational prediction of homodimerization of the A3 adenosine receptor. *J. Mol. Graph. Model.* 25:549-561.

Kirby DA, Boublik JH and Rivier JE (1993) Neuropeptide Y: Y1 and Y2 affinities of the complete series of analogues with single D-residue substitutions. *J. Med. Chem.* 36:3802-3808.

Kirby DA, Koerber SC, May JM, Hagaman C, Cullen MJ, Pelleymounter MA and Rivier JE (1995) Y1 and Y2 receptor selective neuropeptide Y analogues: evidence for a Y1 receptor subclass. *J. Med. Chem.* 38:4579-4586.

Koizumi O, Wilson JD, Grimmelikhuijzen CJ and Westfall JA (1989) Ultrastructural localization of RFamide-like peptides in neuronal dense-cored vesicles in the peduncle of Hydra. *J. Exp. Zool.* 249:17-22.

Kosanam H, Ramagiri S and Dass C (2009) Quantification of endogenous alpha- and gamma-endorphins in rat brain by liquid chromatography-tandem mass spectrometry. *Anal. Biochem.* 392:83-89.

Kota P, Reeves PJ, Rajbhandary UL and Khorana HG (2006) Opsin is present as dimers in COS1 cells: identification of amino acids at the dimeric interface. *Proc. Natl. Acad. Sci. U S A* 103:3054-3059.

Kriegsfeld LJ (2006) Driving reproduction: RFamide peptides behind the wheel. *Horm. Behav.* 50:655-666.

Landry M, Holmberg K, Zhang X and Hokfelt T (2000) Effect of axotomy on expression of NPY, galanin, and NPY Y1 and Y2 receptors in dorsal root ganglia and the superior cervical ganglion studied with double-labeling in situ hybridization and immunohistochemistry. *Exp. Neurol.* 162:361-384.

Larhammar D (1996) Evolution of neuropeptide Y, peptide YY and pancreatic polypeptide. *Regul. Pept.* 62:1-11.

Larhammar D and Salaneck E (2004) Molecular evolution of NPY receptor subtypes. *Neuropeptides* 38:141-151.

Larsen PJ, Sheikh SP, Jakobsen CR, Schwartz TW and Mikkelsen JD (1993) Regional distribution of putative NPY Y(1) receptors and neurons expressing Y(1) messenger RNA in forebrain areas of the rat central nervous system. *Eur. J. Neurosci.* 5:1622-1637.

Lavebratt C, Alpman A, Persson B, Arner P and Hoffstedt J (2006) Common neuropeptide Y2 receptor gene variant is protective against obesity among Swedish men. *Int. J. Obes. (Lond)* 30:453-459.

Leban JJ, Heyer D, Landavazo A, Matthews J, Aulabaugh A and Daniels AJ (1995) Novel modified carboxy terminal fragments of neuropeptide Y with high affinity for Y2-type receptors and potent functional antagonism at a Y1-type receptor. *J. Med. Chem.* 38:1150-1157.

Lecklin A, Lundell I, Salmela S, Mannisto PT, Beck-Sickinger AG and Larhammar D (2003) Agonists for neuropeptide Y receptors Y1 and Y5 stimulate different phases of feeding in guinea pigs. *Br. J. Pharmacol.* 139:1433-1440.

Lee DK, Lanca AJ, Cheng R, Nguyen T, Ji XD, Gobeil F, Jr., Chemtob S, George SR and O'Dowd BF (2004) Agonist-independent nuclear localization of the Apelin, angiotensin AT1, and bradykinin B2 receptors. *J. Biol. Chem.* 279:7901-7908.

Lee K, Li B, Xi X, Suh Y and Martin RJ (2005) Role of neuronal energy status in the regulation of adenosine 5'-monophosphate-activated protein kinase, orexigenic neuropeptides expression, and feeding behavior. *Endocrinology* 146:3-10.

Lerch M, Gafner V, Bader R, Christen B, Folkers G and Zerbe O (2002) Bovine pancreatic polypeptide (bPP) undergoes significant changes in conformation and dynamics upon binding to DPC micelles. *J. Mol. Biol.* 322:1117-1133.

Lerch M, Kamimori H, Folkers G, Aguilar MI, Beck-Sickinger AG and Zerbe O (2005) Strongly altered receptor binding properties in PP and NPY chimeras are accompanied by changes in structure and membrane binding. *Biochemistry* 44:9255-9264.

Lerch M, Mayrhofer M and Zerbe O (2004) Structural similarities of micelle-bound peptide YY (PYY) and neuropeptide Y (NPY) are related to their affinity profiles at the Y receptors. *J. Mol. Biol.* 339:1153-1168.

Li W and Hexum TD (1991) Characterization of neuropeptide Y (NPY) receptors in human hippocampus. *Brain Res.* 553:167-170.

Lundell I, Berglund MM, Starback P, Salaneck E, Gehlert DR and Larhammar D (1997) Cloning and characterization of a novel neuropeptide Y receptor subtype in the zebrafish. *DNA Cell Biol.* 16:1357-1363.

Lundell I, Blomqvist AG, Berglund MM, Schober DA, Johnson D, Statnick MA, Gadski RA, Gehlert DR and Larhammar D (1995) Cloning of a human receptor of the NPY receptor family with high affinity for pancreatic polypeptide and peptide YY. *J. Biol. Chem.* 270:29123-29128.

Lundell I, Boswell T and Larhammar D (2002) Chicken neuropeptide Y-family receptor Y4: a receptor with equal affinity for pancreatic polypeptide, neuropeptide Y and peptide YY. *J. Mol. Endocrinol.* 28:225-235.

Lynch JW, Lemos VS, Bucher B, Stoclet JC and Takeda K (1994) A pertussis toxin-insensitive calcium influx mediated by neuropeptide Y2 receptors in a human neuroblastoma cell line. *J. Biol. Chem.* 269:8226-8233.

Maillo M, Aguilar MB, Lopez-Vera E, Craig AG, Bulaj G, Olivera BM and Heimer de la Cotera EP (2002) Conorfamide, a Conus venom peptide belonging to the RFamide family of neuropeptides. *Toxicon* 40:401-407.

Martin NM, Small CJ, Sajedi A, Patterson M, Ghatei MA and Bloom SR (2004) Pre-obese and obese agouti mice are sensitive to the anorectic effects of peptide YY(3-36) but resistant to ghrelin. *Int. J. Obes. Relat. Metab. Disord* .28:886-893.

McDonald RL, Vaughan PF, Beck-Sickinger AG and Peers C (1995) Inhibition of Ca2+ channel currents in human neuroblastoma (SH-SY5Y) cells by neuropeptide Y and a novel cyclic neuropeptide Y analogue. *Neuropharmacology* 34:1507-1514.

McLean LR, Buck SH and Krstenansky JL (1990) Examination of the role of the amphipathic alpha-helix in the interaction of neuropeptide Y and active cyclic analogues with cell membrane receptors and dimyristoylphosphatidylcholine. *Biochemistry* 29:2016-2022.

Michel MC (1998) Concomitant regulation of Ca2+ mobilization and G13 expression in human erythroleukemia cells. *Eur. J. Pharmacol.* 348:135-141.

Michel MC, Lewejohann K, Farke W, Bischoff A, Feth F and Rascher W (1995) Regulation of NPY/NPY Y1 receptor/G protein system in rat brain cortex. *Am. J. Physiol.* 268:R192-200.

Miner JL, Della-Fera MA, Paterson JA and Baile CA (1989) Lateral cerebroventricular injection of neuropeptide Y stimulates feeding in sheep. *Am. J. Physiol.* 257:R383-387.

Mitchell GC, Wang Q, Ramamoorthy P and Whim MD (2008) A common single nucleotide polymorphism alters the synthesis and secretion of neuropeptide Y. *J. Neurosci.* 28:14428-14434.

Mollereau C, Zajac JM and Roumy M (2007) Staurosporine differentiation of NPFF2 receptor-transfected SH-SY5Y neuroblastoma cells induces selectivity of NPFF activity towards opioid receptors. *Peptides* 28:1125-1128.

Moltz JH and McDonald JK (1985) Neuropeptide Y: direct and indirect action on insulin secretion in the rat. *Peptides* 6:1155-1159.

Moriya R, Mashiko S, Ishihara A, Takahashi T, Murai T, Ito J, Mitobe Y, Oda Z, Iwaasa H, Takehiro F and Kanatani A (2009) Comparison of independent and combined chronic anti-obese effects of NPY Y2 receptor agonist, PYY(3-36), and NPY Y5 receptor antagonist in diet-induced obese mice. *Peptides* 30:1318-1322.

Moulis A (2006) The action of RFamide neuropeptides on molluscs, with special reference to the gastropods Buccinum undatum and Busycon canaliculatum. *Peptides* 27:1153-1165.

Mukhopadhyay S and Ross EM (1999) Rapid GTP binding and hydrolysis by G(q) promoted by receptor and GTPase-activating proteins. *Proc. Natl. Acad. Sci. U S A* 96:9539-9544.

Munson PJ and Rodbard D (1980) LIGAND: a versatile computerized approach for characterization of ligand-binding proteins. *Anal. Biochem.* 107:220-239.

Muroya S, Yada T, Shioda S and Takigawa M (1999) Glucose-sensitive neurons in the rat arcuate nucleus contain neuropeptide Y. *Neurosci. Lett.* 264:113-116.

Nakamura M, Sakanaka C, Aoki Y, Ogasawara H, Tsuji T, Kodama H, Matsumoto T, Shimizu T and Noma M (1995) Identification of two isoforms of mouse neuropeptide Y-Y1 receptor generated by alternative splicing. Isolation, genomic structure, and functional expression of the receptors. *J. Biol. Chem.* 270:30102-30110.

Nakamura M, Yokoyama M, Watanabe H and Matsumoto T (1997) Molecular cloning, organization and localization of the gene for the mouse neuropeptide Y-Y5 receptor. *Biochim. Biophys. Acta* 1328:83-89.

Neubig RR (2007) Missing links: mechanisms of protean agonism. *Mol. Pharmacol.* 71:1200-1202.

Noll T, Hempel A and Piper HM (1996) Neuropeptide Y reduces macro-molecule permeability of coronary endothelial monolayers. *Am. J. Physiol.* 271:H1878-1883.

Nonaka N, Shioda S, Niehoff ML and Banks WA (2003) Characterization of blood-brain barrier permeability to PYY3-36 in the mouse. *J. Pharmacol. Exp. Ther.* 306:948-953.

Nordman S, Ding B, Ostenson CG, Karvestedt L, Brismar K, Efendic S and Gu HF (2005) Leu7Pro polymorphism in the neuropeptide Y (NPY) gene is associated with impaired glucose tolerance and type 2 diabetes in Swedish men. *Exp. Clin. Endocrinol. Diabetes* 113:282-287.

Norenberg W, Bek M, Limberger N, Takeda K and Illes P (1995) Inhibition of nicotinic acetylcholine receptor channels in bovine adrenal chromaffin cells by Y3-type neuropeptide Y receptors via the adenylate cyclase/protein kinase A system. *Naunyn Schmiedebergs Arch. Pharmacol.* 351:337-347.

O'Shea D, Morgan DG, Meeran K, Edwards CM, Turton MD, Choi SJ, Heath MM, Gunn I, Taylor GM, Howard JK, Bloom CI, Small CJ, Haddo O, Ma JJ, Callinan W, Smith DM, Ghatei MA and Bloom SR (1997) Neuropeptide Y induced feeding in the rat is mediated by a novel receptor. *Endocrinology* 138:196-202.

Oksche A, Boese G, Horstmeyer A, Furkert J, Beyermann M, Bienert M and Rosenthal W (2000) Late endosomal/lysosomal targeting and lack of recycling of the ligand-occupied endothelin B receptor. *Mol. Pharmacol.* 57:1104-1113.

Palczewski K, Kumasaka T, Hori T, Behnke CA, Motoshima H, Fox BA, Le Trong I, Teller DC, Okada T, Stenkamp RE, Yamamoto M and Miyano M (2000) Crystal structure of rhodopsin: A G protein-coupled receptor. *Science* 289:739-745.

Papadimitriou MA, Krzemien AA, Hahn PM and Van Vugt DA (2007) Peptide YY(3-36)-induced inhibition of food intake in female monkeys. *Brain Res.* 1175:60-65.

Parker MS, Balasubramaniam A and Parker SL (2012a) On the Segregation of Protein Ionic Residues by Charge Type. *Current Amino Acids* 43:2231-2247.

Parker MS, Berglund MM, Lundell I and Parker SL (2001a) Blockade of pancreatic polypeptide-sensitive neuropeptide Y (NPY) receptors by agonist peptides is prevented by modulators of sodium transport. Implications for receptor signaling and regulation. *Peptides* 22:887-898.

Parker MS, Crowley WR and Parker SL (1996a) Differences in cation sensitivity of ligand binding to Y_1 and Y_2 subtype of neuropeptide Y receptor of rat brain. *Eur. J. Pharmacol.* 318:193-200.

Parker MS, Lundell I and Parker SL (2002a) Internalization of pancreatic polypeptide Y4 receptors: correlation of receptor intake and affinity. *Eur. J. Pharmacol.* 452:279-287.

Parker MS, Lundell I and Parker SL (2002b) Pancreatic polypeptide receptors: affinity, sodium sensitivity and stability of agonist binding. *Peptides* 23:291-302.

Parker MS, Sah R, Balasubramaniam A, Sallee FR, Zerbe O and Parker SL (2011) Non-specific binding and general cross-reactivity of Y receptor agonists are correlated and should importantly depend on their acidic sectors. *Peptides* 32:258-265.

Parker MS, Sah R, Park EA, Sweatman T, Balasubramaniam A, Sallee FR and Parker SL (2009) Oligomerization of the heptahelical G protein coupling receptors: a case for association using transmembrane helices. *Mini Rev. Med. Chem.* 9:329-339.

Parker MS, Sah R and Parker SL (2012b) Surface masking shapes the traffic of the neuropeptide Y Y2 receptor. *Peptides.* 37:40-48

Parker MS, Wang JJ, Fournier A and Parker SL (2000) Upregulation of pancreatic polypeptide-sensitive neuropeptide Y (NPY) receptors in estrogen-induced hypertrophy of the anterior pituitary gland in the Fischer-344 rat. *Mol. Cell Endocrinol* .164:239-249.

Parker SL and Balasubramaniam A (2008) Neuropeptide Y Y2 receptor in health and disease. *Br. J. Pharmacol.* 153:420-431.

Parker SL, Carroll BL, Kalra SP, St-Pierre S, Fournier A and Crowley WR (1996b) Neuropeptide Y Y2 receptors in hypothalamic neuroendocrine areas are up-regulated by estradiol and decreased by progesterone cotreatment in the ovariectomized rat. *Endocrinology* 137:2896-2900.

Parker SL, Kalra SP and Crowley WR (1991) Neuropeptide Y modulates the binding of a gonadotropin-releasing hormone (GnRH) analog to anterior pituitary GnRH receptor sites. *Endocrinology* 128:2309-2316.

Parker SL, Kane JK, Parker MS, Berglund MM, Lundell IA and Li MD (2001b) Cloned neuropeptide Y (NPY) Y1 and pancreatic polypeptide Y4 receptors expressed in Chinese hamster ovary cells show considerable agonist-driven internalization, in contrast to the NPY Y2 receptor. *Eur. J. Biochem.* 268:877-886.

Parker SL and Parker MS (2000) Ligand association with the rabbit kidney and brain Y1, Y2 and Y5-like neuropeptide Y (NPY) receptors shows large subtype-related differences in sensitivity to chaotropic and alkylating agents. *Regul. Pept.* 87:59-72.

Parker SL, Parker MS and Crowley WR (1998a) Characterization of Y1, Y2 and Y5 subtypes of neuropeptide Y (NPY) receptor in rabbit kidney. *Regulatory Peptides* 75/76:127-143.

Parker SL, Parker MS and Crowley WR (1999) Characterization of rabbit kidney and brain pancreatic polypeptide-binding neuropeptide Y receptors: differences with Y1 and Y2 sites in sensitivity to amiloride derivatives affecting sodium transport. *Regul. Pept.* 82:91-102.

Parker SL, Parker MS, Estes MA, Wong YY, Sah R, Sweatman T, Park EA, Balasubramaniam A and Sallee FR (2008a) The neuropeptide Y (NPY) Y2 receptors are largely dimeric in the kidney, but monomeric in the forebrain. *Journal of Receptors and Signal Transduction* 28:245-263.

Parker SL, Parker MS, Kane JK and Berglund MM (2002c) A pool of Y2 neuropeptide Y receptors activated by modifiers of membrane sulfhydryl or cholesterol balance. *Eur. J. Biochem.* 269:2315-2322.

Parker SL, Parker MS, Sah R, Balasubramaniam A and Sallee FR (2007a) Self-regulation of agonist activity at the Y receptors. *Peptides* 28:203-213.

Parker SL, Parker MS, Sah R, Balasubramaniam A and Sallee FR (2008b) Pertussis toxin induces parallel loss of neuropeptide Y Y(1) receptor dimers and G(i) alpha subunit function in CHO cells. *Eur. J. Pharmacol.* 579:13-25.

Parker SL, Parker MS, Sah R, Sallee FR and Balasubramaniam A (2007b) Parallel inactivation of Y2 receptor and G-proteins in CHO cells by pertussis toxin. *Regul. Pept.* 139:128-135.

Parker SL, Parker MS, Sallee FR and Balasubramaniam A (2007c) Oligomerization of neuropeptide Y (NPY) Y2 receptors in CHO cells depends on functional pertussis toxin-sensitive G-proteins. *Regul. Pept.* 144:72-81.

Parker SL, Parker MS, Sweatman T and Crowley WR (1998b) Characterization of G protein and phospholipase C-coupled agonist binding to the Y1 neuropeptide Y receptor in rat brain: sensitivity to G protein activators and inhibitors and to inhibitors of phospholipase C. *J. Pharmacol. Exp.Ther.* 286:382-391.

Parker SL, Parker MS, Wong YY, Sah R, Balasubramaniam A and Sallee F (2008c) Importance of a N-terminal aspartate in the internalization of the neuropeptide Y Y2 receptor. *Eur. J. Pharmacol.* 594:26-31.

Pheng LH, Dumont Y, Fournier A, Chabot JG, Beaudet A and Quirion R (2003) Agonist- and antagonist-induced sequestration/internalization of neuropeptide Y Y1 receptors in HEK293 cells. *Br. J. Pharmacol.* 139:695-704.

Prenner L, Sieben A, Zeller K, Weiser D and Haberlein H (2007) Reduction of high-affinity beta2-adrenergic receptor binding by hyperforin and hyperoside on rat C6 glioblastoma cells measured by fluorescence correlation spectroscopy. *Biochemistry* 46:5106-5113.

Prieto D, Buus C, Mulvany MJ and Nilsson H (1997) Interactions between neuropeptide Y and the adenylate cyclase pathway in rat mesenteric small arteries: role of membrane potential. *J. Physiol.* 502 (Pt 2):281-292.

Protas L, Qu J and Robinson RB (2003) Neuropeptide y: neurotransmitter or trophic factor in the heart? *News Physiol. Sci.* 18:181-185.

Qian J, Colmers WF and Saggau P (1997) Inhibition of synaptic transmission by neuropeptide Y in rat hippocampal area CA1: modulation of presynaptic Ca2+ entry. *J. Neurosci.* 17:8169-8177.

Rakic P (1988) Specification of cerebral cortical areas. *Science* 241:170-176.

Rees AR, Gregoriou M, Johnson P and Garland PB (1984) High affinity epidermal growth factor receptors on the surface of A431 cells have restricted lateral diffusion. *Embo. J* .3:1843-1847.

Rodgers RJ, Holch P and Tallett AJ (2010) Behavioural satiety sequence (BSS): separating wheat from chaff in the behavioural pharmacology of appetite. *Pharmacol. Biochem. Behav.* 97:3-14.

Sah R, Balasubramaniam A, Parker MS, Sallee F and Parker SL (2005) Neuropeptide Y as a partial agonist of the Y1 receptor. *Eur. J. Pharmacol.* 525:60-68.

Sah R, Parker SL, Sheriff S, Eaton K, Balasubramaniam A and Sallee FR (2007) Interaction of NPY compounds with the rat glucocorticoid-induced receptor (GIR) reveals similarity to the NPY-Y2 receptor. *Peptides* 28:302-309.

Sahu A, Sninsky CA, Phelps CP, Dube MG, Kalra PS and Kalra SP (1992) Neuropeptide Y release from the paraventricular nucleus increases in association with hyperphagia in streptozotocin-induced diabetic rats. *Endocrinology* 131:2979-2985.

Sala F, Mulet J, Sala S, Gerber S and Criado M (2005) Charged amino acids of the N-terminal domain are involved in coupling binding and gating in alpha7 nicotinic receptors. *J. Biol. Chem.* 280:6642-6647.

Salaneck E, Holmberg SK, Berglund MM, Boswell T and Larhammar D (2000) Chicken neuropeptide Y receptor Y2: structural and pharmacological differences to mammalian Y2(1). *FEBS Lett.* 484:229-234.

Salaneck E, Larsson TA, Larson ET and Larhammar D (2008) Birth and death of neuropeptide Y receptor genes in relation to the teleost fish tetraploidization. *Gene* 409:61-71.

Sautel M, Rudolf K, Wittneben H, Herzog H, Martinez R, Munoz M, Eberlein W, Engel W, Walker P and Beck-Sickinger AG (1996) Neuropeptide Y and the nonpeptide antagonist BIBP 3226 share an overlapping binding site at the human Y1 receptor. *Mol. Pharmacol.* 50:285-292.

Schillo S, Belusic G, Hartmann K, Franz C, Kuhl B, Brenner-Weiss G, Paulsen R and Huber A (2004) Targeted mutagenesis of the farnesylation site of Drosophila Ggammae disrupts membrane association of the G protein betagamma complex and affects the light sensitivity of the visual system. *J. Biol. Chem.* 279:36309-36316.

Selbie LA, King NV, Dickenson JM and Hill SJ (1997) Role of G-protein beta gamma subunits in the augmentation of P2Y2 (P2U)receptor-stimulated responses by neuropeptide Y Y1 Gi/o-coupled receptors. *Biochem. J.* 328 (Pt 1):153-158.

Serradeil-Le Gal C, Lafontan M, Raufaste D, Marchand J, Pouzet B, Casellas P, Pascal M, Maffrand JP and Le Fur G (2000) Characterization of NPY receptors controlling lipolysis and leptin secretion in human adipocytes. *FEBS Lett.* 475:150-156.

Shamri R, Grabovsky V, Feigelson SW, Dwir O, Van Kooyk Y and Alon R (2002) Chemokine stimulation of lymphocyte alpha 4 integrin avidity but not of leukocyte function-associated antigen-1 avidity to endothelial ligands under shear flow requires cholesterol membrane rafts. *J. Biol. Chem.* 277:40027-40035.

Signoret N, Pelchen-Matthews A, Mack M, Proudfoot AE and Marsh M (2000) Endocytosis and recycling of the HIV coreceptor CCR5. *J. Cell. Biol.* 151:1281-1294.

Silva AP, Kaufmann JE, Vivancos C, Fakan S, Cavadas C, Shaw P, Brunner HR, Vischer U and Grouzmann E (2005) Neuropeptide Y expression, localization and cellular transducing effects in HUVEC. *Biol. Cell* 97:457-467.

Sosulina L, Schwesig G, Seifert G and Pape HC (2008) Neuropeptide Y activates a G-protein-coupled inwardly rectifying potassium current and dampens excitability in the lateral amygdala. *Mol. Cell Neurosci.* 39:491-498.

Stanic D, Mulder J, Watanabe M and Hokfelt T (2011) Characterization of NPY Y2 receptor protein expression in the mouse brain. II. Coexistence with NPY, the Y1 receptor, and other neurotransmitter-related molecules. *J. Comp. Neurol.* 519:1219-1257.

Stanley BG and Leibowitz SF (1985) Neuropeptide Y injected in the paraventricular hypothalamus: a powerful stimulant of feeding behavior. *Proc. Natl. Acad. Sci. U S A* 82:3940-3943.

Starback P, Wraith A, Eriksson H and Larhammar D (2000) Neuropeptide Y receptor gene y6: multiple deaths or resurrections? *Biochem. Biophys. Res. Commun.* 277:264-269.

Statnick MA, Schober DA, Gackenheimer S, Johnson D, Beavers L, Mayne NG, Burnett JP, Gadski R and Gehlert DR (1998) Characterization of the neuropeptide Y5 receptor in the human hypothalamus: a lack of correlation between Y5 mRNA levels and binding sites. *Brain Res.* 810:16-26.

Statnick MA, Schober DA, Mayne NG, Burnett JP and Gehlert DR (1997) Analysis of NPY receptor subtypes in the human frontal cortex reveals abundant Y1 mRNA and binding sites. *Peptides* 18:137-143.

Sun QQ, Huguenard JR and Prince DA (2001) Neuropeptide Y receptors differentially modulate G-protein-activated inwardly rectifying K+ channels and high-voltage-activated Ca2+ channels in rat thalamic neurons. *J. Physiol.* 531:67-79.

Tombola F, Pathak MM and Isacoff EY (2005) How far will you go to sense voltage? *Neuron* 48:719-725.

Travers JB, Herman K and Travers SP (2010) Suppression of third ventricular NPY-elicited feeding following medullary reticular formation infusions of muscimol. *Behav. Neurosci.* 124:225-233.

van den Hoek AM, Voshol PJ, Karnekamp BN, Buijs RM, Romijn JA, Havekes LM and Pijl H (2004) Intracerebroventricular neuropeptide Y infusion precludes inhibition of glucose and VLDL production by insulin. *Diabetes* 53:2529-2534.

Vrang N, Madsen AN, Tang-Christensen M, Hansen G and Larsen PJ (2006) PYY(3-36) reduces food intake and body weight and improves insulin sensitivity in rodent models of diet-induced obesity. *Am. J. Physiol. Regul. Integr. Comp. Physiol* .291:R367-375.

Wahlestedt C, Grundemar L, Hakanson R, Heilig M, Shen GH, Zukowska-Grojec Z and Reis DJ (1990) Neuropeptide Y receptor subtypes, Y1 and Y2. *Ann. N. Y. Acad. Sci.* 611:7-26.

Wang SJ (2005) Activation of neuropeptide Y Y1 receptors inhibits glutamate release through reduction of voltage-dependent Ca2+ entry in the rat cerebral cortex nerve terminals: suppression of this inhibitory effect by the protein kinase C-dependent facilitatory pathway. *Neuroscience* 134:987-1000.

Weinberg DH, Sirinathsinghji DJ, Tan CP, Shiao LL, Morin N, Rigby MR, Heavens RH, Rapoport DR, Bayne ML, Cascieri MA, Strader CD, Linemeyer DL and MacNeil DJ (1996) Cloning and expression of a novel neuropeptide Y receptor. *J. Biol. Chem.* 271:16435-16438.

Weiss JL and Burgoyne RD (2001) Voltage-independent inhibition of P/Q-type Ca2+ channels in adrenal chromaffin cells via a neuronal Ca2+ sensor-1-dependent pathway involves Src family tyrosine kinase. *J. Biol. Chem.* 276:44804-44811.

Whorton MR, Jastrzebska B, Park PS, Fotiadis D, Engel A, Palczewski K and Sunahara RK (2008) Efficient coupling of transducin to monomeric rhodopsin in a phospholipid bilayer. *J. Biol. Chem.* 283:4387-4394.

Wieland HA, Engel W, Eberlein W, Rudolf K and Doods HN (1998) Subtype selectivity of the novel nonpeptide neuropeptide Y Y1 receptor antagonist BIBO 3304 and its effect on feeding in rodents. *Br. J. Pharmacol.* 125:549-555.

Williams TJ, Torres-Reveron A, Chapleau JD and Milner TA (2011) Hormonal regulation of delta opioid receptor immunoreactivity in interneurons and pyramidal cells in the rat hippocampus. *Neurobiol. Learn Mem.* 95:206-220.

Wraith A, Tornsten A, Chardon P, Harbitz I, Chowdhary BP, Andersson L, Lundin LG and Larhammar D (2000) Evolution of the neuropeptide Y receptor family: gene and chromosome duplications deduced from the cloning and mapping of the five receptor subtype genes in pig. *Genome Res.* 10:302-310.

Yasuda D, Okuno T, Yokomizo T, Hori T, Hirota N, Hashidate T, Miyano M, Shimizu T and Nakamura M (2009) Helix 8 of leukotriene B4 type-2 receptor is required for the folding to pass the quality control in the endoplasmic reticulum. *Faseb J.* 23:1470-1481.

Yokobori E, Azuma M, Nishiguchi R, Kang KS, Kamijo M, Uchiyama M and Matsuda K (2012) Neuropeptide Y stimulates food intake in the zebrafish, Danio rerio. *J. Neuroendocrinol.*

Zerbe O, Neumoin A, Mares J, Walser R, Walser R and Zou C (2006) Recognition of neurohormones of the NPY family by their receptors. *J. Recept. Signal. Transduct Res.* 26:487-504.

Zhang H, Roubos EW, Jenks BG and Scheenen WJ (2006) Receptors for neuropeptide Y, gamma-aminobutyric acid and dopamine differentially regulate Ca2+ currents in Xenopus melanotrope cells via the G(i) protein beta/gamma-subunit. *Gen. Comp. Endocrinol.* 145:140-147.

Zhang X, Bao L, Xu ZQ, Kopp J, Arvidsson U, Elde R and Hokfelt T (1994) Localization of neuropeptide Y Y1 receptors in the rat nervous system with special reference to somatic receptors on small dorsal root ganglion neurons. *Proc. Natl. Acad. Sci. U S A* 91:11738-11742.

Zylbergold P, Ramakrishnan N and Hebert T (2010) The role of G proteins in assembly and function of Kir3 inwardly rectifying potassium channels. *Channels (Austin)* 4:411-421.

P